Advances in the Theory of System Decoupling

Rubens Gonçalves Salsa Junior • Fai Ma

Advances in the Theory of System Decoupling

Springer

Rubens Gonçalves Salsa Junior
Aeronautics Institute of Technology
São José dos Campos, São Paulo, Brazil

Fai Ma
University of California at Berkeley
Berkeley, CA, USA

ISBN 978-3-030-60848-4 ISBN 978-3-030-60846-0 (eBook)
https://doi.org/10.1007/978-3-030-60846-0

This Springer imprint is published by the registered company Springer Nature Switzerland AG.
The registered company address is: Gewerbestrasse 11, 6330 Cham, Switzerland

To my wife, Luiza, and my daughter,
Capitu (RGSJ)
To Hsing-Ho, Andrew, and Catherine (FM)

Preface

Linear dynamical systems are characterized by a second-order ordinary differential equation with three real coefficient matrices. The classical decoupling problem in vibrations amounts to converting such systems into a form whose coefficients are diagonal. Coordinate decoupling plays a central role not only in vibrations but also in such diverse areas as numerical linear algebra, quantum mechanics, and mathematical economics. In the past 10 years, the method of modal analysis was generalized to decouple practically all linear systems, including those with non-symmetric coefficient matrices. The extension, called phase synchronization, is based upon compensations for time drifts caused by viscous damping and external excitation. The resulting invertible decoupling transformation is real, non-linear, and time varying.

While phase synchronization could be immensely useful, information on this topic is scattered in various journal articles, requiring the interested readers to resort to multiple references to fully understand the methodology. To make things more complicated, different articles often use different notations. Furthermore, some of these papers might be challenging to read, as both engineers and mathematicians contributed to the development of system decoupling. This gave us our main motivation to write this book: to provide a concise, clear, and consistent exposition of the theory, illustrated didactically by many examples. We hope that graduate students, researchers, and scholars in the fields of engineering, mathematics, and related areas will find this book a didactic and complete source of this topic. We feel that the theory is complete and mature enough for engineering applications. Hence, a secondary objective is to demonstrate the utility of the methodology in solving engineering problems.

The book starts with a survey of common decoupling techniques, summarized and reviewed in Chap. 1. Modal analysis is the classical decoupling methodology in engineering. However, we also include decoupling by similarity transformation and equivalence transformation because these are popular topics in mathematics. The method of phase synchronization is presented in the two subsequent chapters. We have decided to introduce concepts by first considering damped systems because these constitute the bulk of engineering literature and, therefore, should be familiar to readers. General linear systems are considered subsequently in Chap. 3, where systems possessing repeated or defective eigenvalues are also considered. The book is also supplemented by applications of the theory in Chap. 4. While the book surveys recent developments in system decoupling, we believe that we have occasionally obtained a new interpretation or an extension of a known result. The sections on damping ratio and invariant of motion are new and have not been published before. The book is accompanied by a decoupling program consisting of two MATLAB files. The user instructions are provided in Appendix B. The program may be used to solve many of the illustrative examples in this book. Among other things, it is hoped that this book would serve to point to new directions along which further research efforts should be made.

São José dos Campos, SP, Brazil — Rubens Gonçalves Salsa Junior
Berkeley, CA, USA — Fai Ma

Acknowledgments

This project to provide a didactic overview of the method of phase synchronization would not have been possible without the efforts of many collaborators at the University of California, Berkeley. We are very grateful to D.T. Kawano, M. Morzfeld, A. Imam, B.N. Parlett, and G. Leitmann. Part of this book was written while Rubens Gonçalves Salsa Junior was in Berkeley, supported by the CAPES Foundation through a Science without Borders Fellowship, grant number 99999.011952/2013-00. He is also grateful for the support of a Macchi Fellowship at Berkeley.

We are also grateful to the support received from the Acoustic and Dynamics Laboratory at the Aeronautics Institute of Technology, Brazil. We would like to thank D.A. Rade, L.F.F. Rodovalho, and M.B.R. Fernandes for reading an earlier version of this book. Special thanks to T.P. Sales for carefully reviewing the manuscript and providing us with valuable feedback. We also wish to thank Michael Luby and Nandhakumar Sundar at Springer. It was a pleasure working with them.

We hope that this book is free of grammatical and typographical errors. If you find any, please bring them to our attention.

São José dos Campos, SP, Brazil — Rubens Gonçalves Salsa Junior
Berkeley, CA, USA — Fai Ma

Contents

Part I
The Decoupling Problem of Linear Systems

Chapter 1
Linear Systems and Configuration-Space Decoupling Techniques

Look unto the stars to teach us
How the master's thoughts can reach us
Each one follows Newton's math
Silently along its path.

—Albert Einstein

The equation of motion of linear systems is one of the most commonly used equations in science and engineering. It has long been recognized that coordinate coupling in linear systems is a considerable barrier to analysis and design. In this context, it is common for engineers to seek coordinate decoupling techniques. Coordinate decoupling is the process of simultaneously diagonalizing the coefficient matrices of a dynamical system. The main objective of this chapter is to provide an overview of configuration-space decoupling techniques. Three decoupling algorithms are provided, and thirteen examples are supplemented.

1.1 Introduction

The equation of motion of linear systems is one of the most commonly used equations in science and engineering. It amounts to a system of linear second-order ordinary differential equations, combined through three real and square coefficient matrices of the same order. In general the coefficient matrices are not diagonal and the resulting equations are mutually dependent, when the system is termed *coupled*. It has long been recognized that coordinate coupling in linear systems is a considerable barrier to analysis and design. In this regard, the "classical decoupling problem" is concerned with the elimination of coordinate coupling in linear dynamical systems.

Coordinate decoupling is a well-trodden problem that has attracted the attention of many researchers in the past century [1–37]. Decoupling is worthwhile not only because of response calculation but also because of possible system analysis through independent coordinates. For instance, engineers are commonly interested

R. Gonçalves Salsa Junior, F. Ma, *Advances in the Theory of System Decoupling*,
https://doi.org/10.1007/978-3-030-60846-0_1

in physical quantities such as natural frequencies and modes of vibration. In addition, practicing engineers may also be interested in modal reduction (such as the Rayleigh–Ritz method) and computation of modal participation factors or energy distribution among modes.

In 1894 in *The Theory of Sound*, Lord Rayleigh already expounded on the importance of coordinate coupling with the concept of proportional damping [1]. Modal analysis is a time-honored technique for decoupling equations of motion in configuration space [2–16]. It consists of a linear coordinate transformation that decouples the system in real space. However, it is not applicable to most linear systems [17]. In theory, it is always possible to investigate linear system with the Hamiltonian approach, whereby the second-order equation is recast into a first-order form of dimension two times greater [18–20]. But this method has never appealed to engineers mainly because there is serious absence of physical insight in tackling a first-order equation recast from the Lagrangian formulation.

Ideally, it would be preferable if classical modal analysis could somehow be modified to treat general linear systems. Over the years, two techniques have emerged for decoupling general linear systems in configuration space as extensions of modal analysis. The first technique involves decoupling by a common similarity transformation [21–25]. The second technique involves simultaneous reduction of the coefficient matrices to diagonal forms by equivalence transformation [26]. Both techniques suffer from mathematical restrictions. Most importantly, the decoupled systems may have complex coefficients. Therefore, these techniques did not find their ways into applications.

In 2009, Ma et al. [27] proposed an extension of modal analysis to decouple damped system in oscillatory free vibration. It was termed the method of phase synchronization because its main idea is based on synchronization of phase shifts caused by viscous damping. In that light, each damped mode is converted into a classically damped mode, which can then be associated with a real decoupled system through a real decoupling transformation in configuration space. Later, it was also shown that the same methodology could be carried over to non-oscillatory free vibration [28]. In order to accomplish that, it was necessary to use the concept of vibration with imaginary frequency. Every formula established for oscillatory free vibration could then be suitably applied to purely non-oscillatory free vibration. Subsequently, phase synchronization operates in the same way as for oscillatory damped modes, amounting to introducing complex phase shifts to each mode so that a classically damped mode is generated. In this same paper, it was also shown how decoupling of forced systems is incorporated in the methodology of phase synchronization. Contrary to the modal transformation, the decoupling transformation by phase synchronization must account for additional and not necessarily constant time shifts caused by the external force, thus it is dependent on it. The decoupling transformation in configuration space is real, time dependent, and non-linear [29]. In 2011, Morzfeld et al. [30] showed how to compute the time shifts for systems whose coefficient matrices lacked predefined structure, such as symmetry. The lack of symmetry and positive-definiteness would allow, for example, exponentially increasing modes of vibration. Therefore, phase synchronization was shown to be applicable not only to damped systems but also to linear systems in general. While the works in [27–30] considered systems with distinct eigenvalues, Kawano et al. [31] showed that the methodology could still be applied to systems possessing repeated or defective eigenvalues. In this case, the different multiplicities of the eigenvalues imply that not every system may be decoupled isospectrally [32]. While the invertibility of the acceleration coefficient will be assumed here, the works in [32] and [33] dealt with the application of phase synchronization for the case of a singular leading coefficient.

The main objectives of this book are two-fold. First, it is intended to present a complete overview of the method of phase synchronization. Several illustrative examples are provided to reinforce the important aspects of the theory. The other objective is to delineate how the method can be applied to solve engineering problems.

1.1.1 Chapter Organization and Notation

This chapter aims at reviewing classical decoupling techniques and their limitations. It is organized as follows. The form of the equation of motion of linear systems is discussed in Sect. 1.2. The main properties of the associated non-linear eigenvalue problem are presented in Sect. 1.2.3. This is followed by a review of classical decoupling techniques. In Sect. 1.3 a brief review of modal analysis is presented. Decoupling by similarity transformation is addressed in Sect. 1.4 and equivalence transformations in Sect. 1.5.

Before proceeding, it is convenient to introduce the notation used in this text. It should be noted that these notes were written by engineers, and special emphasis is given to mechanical and structural systems notation. Scalars are presented by lowercase letters as well as Greek symbols. Even though a real number is also a complex number, complex numbers will be taken as those with a nonzero imaginary part. Vectors are represented by lowercase bold symbols, while matrices correspond to uppercase bold symbols. The complex conjugate of λ, $\mathbf{v}$, and $\mathbf{A}$ are denoted by $\bar{\lambda}$, $\bar{\mathbf{v}}$, and $\bar{\mathbf{A}}$, respectively. The transpose of $\mathbf{v}$ and $\mathbf{A}$ are $\mathbf{v}^T$ and $\mathbf{A}^T$, respectively. The complex conjugate transpose of $\mathbf{v}$ and $\mathbf{A}$ are $\mathbf{v}^*$ and $\mathbf{A}^*$, respectively. The inverse of a matrix $\mathbf{A}$ is $\mathbf{A}^{-1}$, and the determinant of $\mathbf{A}$ is $\det(\mathbf{A})$. Appendix A furnishes a complete list of commonly used symbols and their meanings.

1.2 Coordinate Coupling in the Equation of Motion

Any set of independent variables that can be used to describe a mechanical system completely is a set of independent generalized coordinates for the system, the number of which is equal to the number of degrees of freedom [38]. Suppose a mechanical system is specified by the independent generalized coordinates $q_1,\ q_2,\ \ldots,\ q_n$. The n-dimensional real space spanned by the vector

$$\mathbf{q} = \begin{bmatrix} q_1 \\ q_2 \\ \vdots \\ q_n \end{bmatrix} \tag{1.1}$$

is called the configuration space, and the solution curve in configuration space is called a dynamical path. In classical formulation of dynamics, a holonomic system specified by generalized coordinates q_j $(j = 1,\ 2,\ \ldots,\ n)$ is governed by a set of Lagrange's equations:

$$\frac{d}{dt}\left(\frac{\partial L}{\partial \dot{q}_j}\right) - \frac{\partial L}{\partial q_j} = Q_j, \quad j = 1,\ 2,\ \ldots,\ n. \tag{1.2}$$

In the above differential equations, the scalar $L(\mathbf{q},\ \dot{\mathbf{q}},\ t)$ is the system Lagrangian, Q_j is the generalized non-conservative force associated with the coordinate q_j, and overdots represent differentiation with respect to time. It must be kept in mind that conservative forces are derivable from a potential function, and as such they are accounted for in the Lagrangian, of which the potential energy is a part. In traditional applications, Q_j $(j = 1,\ 2,\ \ldots,\ n)$ depend on $\mathbf{q}$, $\dot{\mathbf{q}}$, and t.

Upon simplification, the equations of motion can be cast in the form [39–41]:

$$\mathbf{M}(\mathbf{q},\ t)\,\ddot{\mathbf{q}} + \mathbf{f}(\mathbf{q},\ \dot{\mathbf{q}},\ t) = \mathbf{0}, \tag{1.3}$$

where $\mathbf{M}(\mathbf{q},\ t)$ is a symmetric and positive-definite $n \times n$ matrix function and $\mathbf{f}(\mathbf{q},\ \dot{\mathbf{q}},\ t)$ is a non-linear n-dimensional vector function. The above equation describes the non-linear motion of multi-degree-of-freedom systems. It is worth noting that Lagrange's equations are linear in the accelerations.

Perhaps an alternative viewpoint may be brought up to understand the form of Eq. (1.3). A single-degree-of-freedom system is such that $n = 1$ and all vectorial quantities in (1.3) are real scalars, in which case Eq. (1.3) takes the form:

$$\ddot{q} + f(q,\ \dot{q},\ t) = 0. \tag{1.4}$$

Such a non-linear equation may describe the motion of a particle. The motion of a particle is completely specified by its position q and velocity $\dot{q}$ at any time t or, in other words, by its state $(q,\ \dot{q})$. In the absence of external forces, a particle continues in its state of rest or uniform motion in a straight line. When acted upon by an impressed force, an acceleration $\ddot{q}$ is generated. The impressed force $f(q,\ \dot{q},\ t)$ can only depend on the state $(q,\ \dot{q})$ and time t; it cannot depend on $\ddot{q}$, which is generated as a consequence.

In applications, engineering systems of practical significance are usually assumed to obey linear differential equations as a first approximation. It is then customary to linearize Eq. (1.3) about the origin $\mathbf{q} = \mathbf{0}$, which can be assumed to be an equilibrium point so that $\dot{\mathbf{q}} = \mathbf{0}$ [42–44]. The generalized Taylor expansion about this equilibrium point leads to the approximations [45]:

$$\mathbf{M}(\mathbf{q},\ t) = \mathbf{M}(\mathbf{0},\ t) + \ldots \cong \mathbf{M}(t) \tag{1.5}$$

and

$$\mathbf{f}(\mathbf{q},\ \dot{\mathbf{q}},\ t) = \mathbf{f}(\mathbf{0},\ \mathbf{0},\ t) + \mathbf{A}_0(t)\,\mathbf{q} + \mathbf{A}_1(t)\,\dot{\mathbf{q}} + \ldots \cong \mathbf{f}(t) + \mathbf{A}_0(t)\,\mathbf{q} + \mathbf{A}_1(t)\,\dot{\mathbf{q}} \tag{1.6}$$

up to first-order terms. In this expansion, $\mathbf{A}_1(t)$ and $\mathbf{A}_0(t)$ are real square matrices of order n. Upon linearization, the equation of motion (1.3) takes the form:

$$\mathbf{M}(t)\,\ddot{\mathbf{q}} + \mathbf{A}_1(t)\,\dot{\mathbf{q}} + \mathbf{A}_0(t)\,\mathbf{q} = \mathbf{f}(t), \tag{1.7}$$

where the excitation $-\mathbf{f}(t)$ has been replaced, without loss of generality, by $\mathbf{f}(t)$.

A linear system is time invariant if and only if it possesses constant coefficients. This condition arises when the holonomic constraints are time independent. Under the assumption of time invariance, the equations of motion take the form:

$$\mathbf{M}\ddot{\mathbf{q}} + \mathbf{A}_1\dot{\mathbf{q}} + \mathbf{A}_0\mathbf{q} = \mathbf{f}(t), \tag{1.8}$$

where $\mathbf{M}$, $\mathbf{A}_1$, and $\mathbf{A}_0$ are constant real square matrices of order n. The mass matrix $\mathbf{M} = \mathbf{M}^T$ is symmetric and positive-definite, but $\mathbf{A}_1$ and $\mathbf{A}_0$ are arbitrary. However, a real square matrix can always be decomposed as the sum of a real symmetric matrix and a real skew-symmetric matrix. In that view, a linear time-invariant system may be expressed as

$$\mathbf{M}\ddot{\mathbf{q}} + (\mathbf{C} + \mathbf{G})\dot{\mathbf{q}} + (\mathbf{K} + \mathbf{H})\mathbf{q} = \mathbf{f}(t). \tag{1.9}$$

The coefficient matrix of velocity is decomposed as $\mathbf{A}_1 = \mathbf{C} + \mathbf{G}$, where $\mathbf{C} = \mathbf{C}^T$ is a damping matrix and $\mathbf{G} = -\mathbf{G}^T$ is a gyroscopic matrix. Likewise, the coefficient matrix of displacement can be decomposed as $\mathbf{A}_0 = \mathbf{K} + \mathbf{H}$, where $\mathbf{K} = \mathbf{K}^T$ is a stiffness matrix and $\mathbf{H} = -\mathbf{H}^T$ is a circulatory matrix. Gyroscopic forces do not dissipate energy, while circulatory forces account for constraint damping. A linear time-invariant system in free vibration is conservative if the damping and circulatory matrices are zero.

For a large majority of systems, including all non-rotating structures, the gyroscopic and circulatory matrices are zero and the familiar equation:

$$\mathbf{M}\ddot{\mathbf{q}} + \mathbf{C}\dot{\mathbf{q}} + \mathbf{K}\mathbf{q} = \mathbf{f}(t) \tag{1.10}$$

is obtained. Passive or damped linear systems possess symmetric and positive-definite matrix coefficients **M**, **C**, and **K**, a condition that follows from physical consideration and ensures stability. There should be no denying that the bulk of existing literature on linear vibration and structural dynamics deals implicitly or explicitly with passive systems [2–16].

Throughout the previous development, the coefficient matrix of acceleration **M** remains symmetric. Specific engineering examples involving a non-symmetric coefficient matrix of acceleration have been given by Schmitz [46] as well as Soom and Kim [47]. In this sense, the most general second-order time-invariant linear system takes the form [48]:

$$\mathbf{A}_2\ddot{\mathbf{q}} + \mathbf{A}_1\dot{\mathbf{q}} + \mathbf{A}_0\mathbf{q} = \mathbf{f}(t), \tag{1.11}$$

where the coefficient matrix of acceleration $\mathbf{A}_2$ is no longer symmetric. Eq. (1.11) contains Eqs. (1.8) and (1.10) as particular cases, and for this reason it will be used as the prototype equation of a linear system, representing the so-called linear non-conservative systems [49]. However, $\mathbf{A}_2$ is assumed non-singular. This assumption is not unduly restrictive, as it might be possible to initially decrease the number of degrees of freedom to ensure that $\mathbf{A}_2$ is non-singular [4].

In traditional applications, equations of the above type arise mostly in the area of vehicle dynamics [26]. The use of control devices in structures, however, has permitted linear non-conservative systems to manifest on a widespread scale. The various ways in which control gains can be embedded in the coefficients and the excitation of system (1.11) are discussed by Soong [50].

In general, the j-th component equation of (1.11) involves not only q_j and its derivatives but also other coordinates and their derivatives as well, because the coefficient matrices are not necessarily diagonal. Thus, the vector differential equation of a multi-degree-of-freedom system (1.11) is said to be coupled.

The "classical decoupling problem" is concerned with the elimination of coordinate coupling in linear dynamical systems. Mathematically, it is equivalent to the problem of simultaneous conversion of $\mathbf{A}_2$, $\mathbf{A}_1$, and $\mathbf{A}_0$ into diagonal forms. It is the process of simultaneously diagonalizing the coefficient matrices in system (1.11), which is then transformed into

$$\ddot{\mathbf{p}} + \mathbf{D}\dot{\mathbf{p}} + \boldsymbol{\Omega}\mathbf{p} = \mathbf{g}(t), \tag{1.12}$$

where the matrices **D** and $\boldsymbol{\Omega}$ are diagonal. When simultaneous diagonalization applies, a linear non-conservative system is completely decoupled.

1.2.1 Example 1.1

This example illustrates the process of obtaining the linear equations of motion. The motion of the double pendulum illustrated in Fig. 1.1 is characterized by the vector

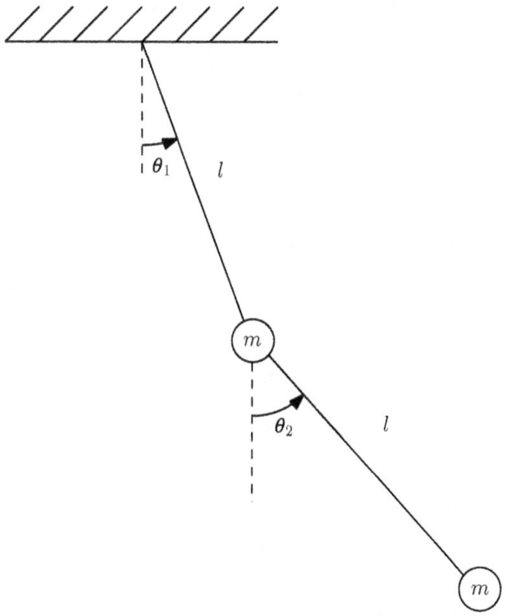

Fig. 1.1 A double pendulum that consists of two identical pendulums, each with mass m and length l. The generalized coordinates are the angular coordinates θ_1 and θ_2

of generalized coordinates $\mathbf{q} = [\theta_1 \quad \theta_2]^T$. The equation of motion is of the form (1.3) for which [40]

$$\mathbf{M}(\mathbf{q},\ t) = \begin{bmatrix} 2ml^2 & ml^2\cos(\theta_2 - \theta_1) \\ ml^2\cos(\theta_2 - \theta_1) & ml^2 \end{bmatrix} \tag{1.13}$$

and

$$\mathbf{f}(\mathbf{q},\ \dot{\mathbf{q}},\ t) = \begin{bmatrix} -ml^2\dot{\theta}_2^2\sin(\theta_2 - \theta_1) + 2mgl\sin\theta_1 \\ \\ ml^2\dot{\theta}_1^2\sin(\theta_2 - \theta_1) + mgl\sin\theta_2 \end{bmatrix}. \tag{1.14}$$

The equation of small oscillations about the equilibrium position $\theta_1 = \theta_2 = 0$ is obtained by the Taylor expansion of each of these terms. The coefficient $\mathbf{M}(\mathbf{q},\ t)$ is simply

$$\mathbf{M}(\mathbf{q},\ t) \cong \mathbf{M}(\mathbf{0},\ t) = ml^2\begin{bmatrix} 2 & 1 \\ 1 & 1 \end{bmatrix}. \tag{1.15}$$

The coefficient $\mathbf{f}(\mathbf{q},\ \dot{\mathbf{q}},\ t)$ becomes

$$\mathbf{f}(\mathbf{q},\ \dot{\mathbf{q}},\ t) \cong \mathbf{f}(t) + \mathbf{A}_0(t)\,\mathbf{q} + \mathbf{A}_1(t)\,\dot{\mathbf{q}}$$

$$= \begin{bmatrix} \frac{\partial f_1}{\partial \theta_1}(\mathbf{0},\mathbf{0},t) & \frac{\partial f_1}{\partial \theta_2}(\mathbf{0},\mathbf{0},t) \\ \frac{\partial f_2}{\partial \theta_1}(\mathbf{0},\mathbf{0},t) & \frac{\partial f_2}{\partial \theta_2}(\mathbf{0},\mathbf{0},t) \end{bmatrix} \mathbf{q} = mgl \begin{bmatrix} 2 & 0 \\ 0 & 1 \end{bmatrix} \begin{bmatrix} \theta_1 \\ \theta_2 \end{bmatrix}. \tag{1.16}$$

Therefore the linearized equation of motion is

$$\underbrace{\begin{bmatrix} 2 & 1 \\ 1 & 1 \end{bmatrix}}_{\mathbf{M}=\mathbf{M}^T} \begin{bmatrix} \ddot{\theta}_1 \\ \ddot{\theta}_2 \end{bmatrix} + \underbrace{\frac{g}{l} \begin{bmatrix} 2 & 0 \\ 0 & 1 \end{bmatrix}}_{\mathbf{K}=\mathbf{K}^T} \begin{bmatrix} \theta_1 \\ \theta_2 \end{bmatrix} = \underbrace{\begin{bmatrix} 0 \\ 0 \end{bmatrix}}_{\mathbf{f}(t)}. \tag{1.17}$$

1.2.2 Other Applications in Science and Engineering

The above development has been directed toward the dynamics of discrete mechanical systems. However, second-order linear differential equations of the form (1.11) also stem from discretization techniques for partial differential equations. In mechanical engineering, these may arise from the governing equations of motion of elastic rods, beams, or shafts [51]. There are many numerical techniques available for discretization [52]. In particular, the finite differences method [53] and the finite elements method [54, 55] may produce equations of the form (1.10) with symmetric and banded matrix coefficients.

Besides mechanical engineering applications, these types of equations are also abundant in electrical engineering. They may be used to analyze the behavior of second-order circuits, like the one depicted in Fig. 1.2. This electrical circuit is described by an equation of the form (1.10) with symmetric coefficients given by

$$\mathbf{M} = \begin{bmatrix} L_1 & 0 \\ 0 & L_2 \end{bmatrix}, \tag{1.18}$$

$$\mathbf{C} = \begin{bmatrix} R_1 + R_2 & -R_2 \\ -R_2 & R_2 + R_3 \end{bmatrix}, \tag{1.19}$$

and

$$\mathbf{K} = \begin{bmatrix} 1/C_1 + 1/C_2 & -1/C_2 \\ -1/C_2 & 1/C_2 + 1/C_3 \end{bmatrix}. \tag{1.20}$$

It is thus safe to say that the equation of motion of linear dynamical systems (1.11) is one of the most used equations in science and engineering. Due to the wide range of applications, units are not emphasized in the illustrative examples.

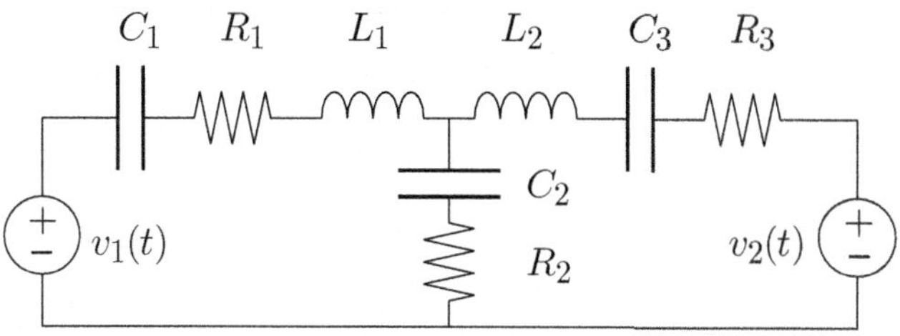

Fig. 1.2 An electrical circuit characterized by second-order differential equations of the form (1.10). L_1 and L_2 are inductances, C_1, C_2, and C_3 are capacitances, and R_1, R_2, and R_3 are resistances

1.2.3 Free Vibration Response and the Quadratic Eigenvalue Problem

Associated with system (1.11) is the quadratic (non-linear) eigenvalue problem:

$$(\lambda^2 \mathbf{A}_2 + \lambda \mathbf{A}_1 + \mathbf{A}_0)\mathbf{v} = \mathbf{0}, \tag{1.21}$$

where $\lambda \in \mathbb{C}$ is an eigenvalue and $\mathbf{v} \in \mathbb{C}^n$ is an n-dimensional eigenvector. The matrix polynomial:

$$\mathbf{Q}(\lambda) = \lambda^2 \mathbf{A}_2 + \lambda \mathbf{A}_1 + \mathbf{A}_0 \tag{1.22}$$

is often called a λ-matrix [56–59]. The eigenvalues are the roots of the characteristic polynomial:

$$\det \mathbf{Q}(\lambda) = \det(\lambda^2 \mathbf{A}_2 + \lambda \mathbf{A}_1 + \mathbf{A}_0) = 0, \tag{1.23}$$

which is of degree $2n$ when the determinant is expanded. The fundamental theorem of algebra states that a polynomial equation of degree $2n$ has $2n$ roots, counting multiplicities of repeated roots. An explicit formula for the roots of polynomial equations of high degree has not been found. Hence determination of eigenvalues is iterative in nature for large n [60, 61].

Solution of (1.23) yields $2n$ eigenvalues λ_j $(j = 1, 2, \ldots, 2n)$, and the corresponding eigenvectors are in the null space of $\mathbf{Q}(\lambda_j)$, each is determined at most up to a multiplicative constant. The set $\{\lambda \in \mathbb{C} : \det \mathbf{Q}(\lambda) = 0\}$ formed by the eigenvalues is the spectrum of system (1.11). Since the coefficient $\mathbf{A}_2$ is invertible, all eigenvalues are finite [59]. Clearly, the spectrum is symmetric with respect to the real axis of the complex plane, and therefore the eigenvalues and eigenvectors are either real or occur in complex conjugate pairs. It is assumed that, out of the $2n$ eigensolutions, $2n_c$ are complex conjugate pairs and the remaining $2n_r = 2(n - n_c)$ are real. An eigenvector of a real eigenvalue can always be chosen real.

The algebraic multiplicity β of an eigenvalue λ is the number of times that it is repeated, i.e., the order of the corresponding zero in (1.23). The geometric multiplicity γ of λ is the number of independent eigenvectors associated with λ, i.e., the dimension of the null space of $\mathbf{Q}(\lambda)$. An eigenvalue is simple if it occurs only once; the algebraic multiplicity and the geometric multiplicity of a simple eigenvalue are both equal to one: $\beta = \gamma = 1$.

A repeated eigenvalue is either semi-simple or defective. A repeated eigenvalue is semi-simple or non-defective if its algebraic and geometric multiplicities coincide: $\beta = \gamma$. This means a semi-simple eigenvalue must be repeated at most n times. Associated with a semi-simple eigenvalue λ with algebraic multiplicity $\beta \leq n$ are β linearly independent eigenvectors $\mathbf{v}_j \neq \mathbf{0}$ $(j = 1, 2, \ldots, \beta)$ in the null space of $\mathbf{Q}(\lambda)$. Consequently, a simple eigenvalue has a single eigenvector $\mathbf{v} \neq \mathbf{0}$. When the linear system only has simple or semi-simple eigenvalues, it is termed non-defective.

The eigenvector associated with a simple eigenvalue can be determined up to a multiplicative constant. In the case of a repeated eigenvalue, any linear combination of the associated eigenvectors is an eigenvector. In this case, the eigenvectors can only be determined within the linear subspace spanned by the linear combination of the independent eigenvectors.

Experience indicates that an eigenvalue problem possessing physical significance is invariably non-defective. Perhaps an alternative viewpoint of the significance of defective eigenvalue problems should be brought up. It is recognized that the probability of a randomly chosen (uniformly distributed in a bounded continuous domain) square matrix being singular is zero. Since a square matrix is singular if and only if its column vectors are dependent, the probability that a set of randomly chosen column vectors being dependent, therefore, is zero. The eigenvectors of Eq. (1.21) are continuous functions of $\mathbf{A}_2$, $\mathbf{A}_1$, and $\mathbf{A}_0$ (except on a set of measure zero). If these matrices are random (uniformly distributed), the resulting eigenvectors are also random. Therefore, loosely speaking, the probability of the eigenvalue problem (1.21) being defective is zero. In this sense, almost all linear systems are characterized by distinct eigenvalues.

While defective systems do not typically arise in practical applications, they have received attention from a number of authors [31, 32, 56–58, 62, 63]. For this reason, defective eigenvalues will be addressed. When an eigenvalue λ_0 occurs β times and a full complement of β independent eigenvectors cannot be found, that eigenvalue is defective. In this case, the $\gamma < \beta$ eigenvectors must be supplemented by $\beta - \gamma$ generalized eigenvectors. A semi-simple eigenvalue is an eigenvalue for which there is no generalized eigenvector.

In a Jordan canonical form, the size of each Jordan block is termed its partial multiplicity. An eigenvalue might have several partial multiplicities, each denoted as μ_k. The algebraic multiplicity is the sum of its partial multiplicities, while the number of partial multiplicities is the geometric multiplicity. A simple eigenvalue has unit partial multiplicity, and a semi-simple eigenvalue repeated β times has β unit partial multiplicities.

Associated with each partial multiplicity $\mu_k > 1$ of a defective eigenvalue λ_0 is a *Jordan chain* $\mathbf{v}_k^l$ $(l = 1, 2, \ldots, \mu_k)$ of length μ_k, where $\mathbf{v}_k^1 \neq \mathbf{0}$ is an eigenvector and $\mathbf{v}_k^l$ $(l = 2, 3, \ldots, \mu_k)$ are generalized eigenvectors satisfying [57]

$$\begin{bmatrix} \mathbf{Q}(\lambda_0) & \mathbf{0} & \cdots & \mathbf{0} \\ \mathbf{Q}^{(1)}(\lambda_0) & \mathbf{Q}(\lambda_0) & \cdots & \mathbf{0} \\ \vdots & \vdots & \ddots & \vdots \\ \dfrac{\mathbf{Q}^{(\mu_k-1)}(\lambda_0)}{(\mu_k-1)!} & \dfrac{\mathbf{Q}^{(\mu_k-2)}(\lambda_0)}{(\mu_k-2)!} & \cdots & \mathbf{Q}(\lambda_0) \end{bmatrix} \begin{bmatrix} \mathbf{v}_k^1 \\ \mathbf{v}_k^2 \\ \vdots \\ \mathbf{v}_k^{\mu_k} \end{bmatrix} = \mathbf{0}. \tag{1.24}$$

In the above sequence, $\mathbf{Q}^{(j)}(\lambda_0)$ denotes the j-th derivative of $\mathbf{Q}$ with respect to λ that is evaluated at the defective eigenvalue λ_0. In contrast to the Jordan chain of a single matrix, the generalized eigenvectors need not be linearly independent. Furthermore, it is possible for a generalized eigenvector to be the zero vector [59]. If an eigenvalue is repeated more than n times, it is necessarily defective.

The general solution of system (1.11) is obtained by adding the homogeneous solution to a particular solution. Regardless of the nature of its spectrum, the free motion response has the general form [57, 58]:

$$\mathbf{q}(t) = \mathbf{V}_q e^{\mathbf{J}_q t} \mathbf{c}, \tag{1.25}$$

for which $\mathbf{V}_q$ is an $n \times 2n$ matrix whose columns correspond to the associated eigenvectors and generalized eigenvectors, and $\mathbf{c}$ is a $2n$-dimensional column vector of coefficients determined from the initial conditions $\mathbf{q}(0)$ and $\dot{\mathbf{q}}(0)$. The corresponding $2n \times 2n$ Jordan matrix $\mathbf{J}_q$ is a block diagonal matrix with the eigenvalues in its diagonal, and the form of each block depends on the associated eigenvalue. Each simple eigenvalue corresponds to a 1×1 block, corresponding to diagonal entries in $\mathbf{J}_q$. A semi-simple eigenvalue repeated β times corresponds to β diagonal entries in $\mathbf{J}_q$. A defective eigenvalue λ_0 is associated with as many blocks as the number of partial multiplicities. Each block associated with a defective eigenvalue λ_0 with a partial multiplicity μ_k has dimension $\mu_k \times \mu_k$ of the form:

$$\begin{bmatrix} \lambda_0 & 1 & & & \\ & \lambda_0 & 1 & & \\ & & \lambda_0 & & \\ & & & \ddots & 1 \\ & & & & \lambda_0 \end{bmatrix}, \tag{1.26}$$

where all remaining entries are equal to zero. The nonzero off-diagonal entries result in powers of t in the solution (1.25) because of the nature of the matrix exponential

$e^{\mathbf{J}_q t}$. In the event that the $2n$ eigenvalues of (1.21) are distinct, $\mathbf{J}_q$ simplifies to the diagonal form

$$\mathbf{J}_q = \begin{bmatrix} \lambda_1 & & & \\ & \lambda_2 & & \\ & & \ddots & \\ & & & \lambda_{2n} \end{bmatrix}, \tag{1.27}$$

$\mathbf{V}_q$ becomes a matrix of eigenvectors

$$\mathbf{V}_q = \begin{bmatrix} \mathbf{v}_1 & \mathbf{v}_2 & \cdots & \mathbf{v}_{2n} \end{bmatrix} \tag{1.28}$$

and response (1.25) can be written as [57, 58]

$$\mathbf{q}(t) = \sum_{j=1}^{2n} c_j \mathbf{v}_j e^{\lambda_j t}. \tag{1.29}$$

A linear system (1.11) is stable if its solution is bounded or decays over time. The homogeneous solution is stable if $\mathrm{Re}[\lambda_j] \leq 0$ for simple eigenvalues and $\mathrm{Re}[\lambda_j] < 0$ for repeated eigenvalues. The homogeneous system is asymptotically stable if $\mathrm{Re}[\lambda_j] < 0$ for every eigenvalue. This is true for a passive system (1.10). The system is unstable if $\mathrm{Re}[\lambda_j] > 0$ for any eigenvalue.

1.2.3.1 Example 1.2

A three-degree-of-freedom system of the form (1.10) has coefficients:

$$\mathbf{M} = \mathbf{I},\ \mathbf{C} = \begin{bmatrix} 4 & -1 & 0 \\ -1 & 2 & -1 \\ 0 & -1 & 2 \end{bmatrix},\ \mathbf{K} = \begin{bmatrix} 3 & -1 & 0 \\ -1 & 2 & -1 \\ 0 & -1 & 1 \end{bmatrix}. \tag{1.30}$$

Solution of the characteristic polynomial (1.23) reveals that the system possesses two simple real eigenvalues $\lambda_1 = -2 - \sqrt{2}$ and $\lambda_2 = -2 + \sqrt{2}$ with eigenvectors

$$\mathbf{v}_1 = \begin{bmatrix} (1+\sqrt{2})^2 \\ -(1+\sqrt{2}) \\ 1 \end{bmatrix},\ \mathbf{v}_2 = \begin{bmatrix} (-1+\sqrt{2})^2 \\ -1+\sqrt{2} \\ 1 \end{bmatrix}, \tag{1.31}$$

respectively. The other eigenvalue $\lambda_3 = -1$ is repeated four times, which means it is defective. It has two independent eigenvectors:

$$\mathbf{v}_3 = \begin{bmatrix} 1 \\ 0 \\ 0 \end{bmatrix}, \; \mathbf{v}_4 = \begin{bmatrix} 0 \\ 0 \\ 2 \end{bmatrix}. \tag{1.32}$$

The eigenvector $\mathbf{v}_4 = \mathbf{v}_4^1$ is supplemented by two generalized eigenvectors

$$\mathbf{v}_4^2 = \begin{bmatrix} 1 \\ 2 \\ 1 \end{bmatrix}, \; \mathbf{v}_4^3 = \begin{bmatrix} 0 \\ 2 \\ 0 \end{bmatrix} \tag{1.33}$$

obtained from (1.24). This means

$$\begin{aligned} \mathbf{V}_q &= \begin{bmatrix} \mathbf{v}_1 & \mathbf{v}_2 & \mathbf{v}_3 & \mathbf{v}_4^1 & \mathbf{v}_4^2 & \mathbf{v}_4^3 \end{bmatrix} \\ &= \begin{bmatrix} (1+\sqrt{2})^2 & (-1+\sqrt{2})^2 & 1 & 0 & 1 & 0 \\ -(1+\sqrt{2}) & -1+\sqrt{2} & 0 & 0 & 2 & 2 \\ 1 & 1 & 0 & 2 & 1 & 0 \end{bmatrix} \end{aligned} \tag{1.34}$$

and the corresponding Jordan matrix is

$$\begin{aligned} \mathbf{J}_q &= \begin{bmatrix} \lambda_1 & 0 & 0 & 0 & 0 & 0 \\ 0 & \lambda_2 & 0 & 0 & 0 & 0 \\ 0 & 0 & \lambda_3 & 0 & 0 & 0 \\ 0 & 0 & 0 & \lambda_3 & 1 & 0 \\ 0 & 0 & 0 & 0 & \lambda_3 & 1 \\ 0 & 0 & 0 & 0 & 0 & \lambda_3 \end{bmatrix} \\ &= \begin{bmatrix} -2-\sqrt{2} & 0 & 0 & 0 & 0 & 0 \\ 0 & -2+\sqrt{2} & 0 & 0 & 0 & 0 \\ 0 & 0 & -1 & 0 & 0 & 0 \\ 0 & 0 & 0 & -1 & 1 & 0 \\ 0 & 0 & 0 & 0 & -1 & 1 \\ 0 & 0 & 0 & 0 & 0 & -1 \end{bmatrix}. \end{aligned} \tag{1.35}$$

The homogeneous response is given by

$$\mathbf{q}(t) = \mathbf{V}_q \begin{bmatrix} e^{(-2-\sqrt{2})t} & 0 & 0 & 0 & 0 & 0 \\ 0 & e^{(-2+\sqrt{2})t} & 0 & 0 & 0 & 0 \\ 0 & 0 & e^{-t} & 0 & 0 & 0 \\ 0 & 0 & 0 & e^{-t} & te^{-t} & t^2e^{-t}/2 \\ 0 & 0 & 0 & 0 & e^{-t} & te^{-t} \\ 0 & 0 & 0 & 0 & 0 & e^{-t} \end{bmatrix} \mathbf{c}, \tag{1.36}$$

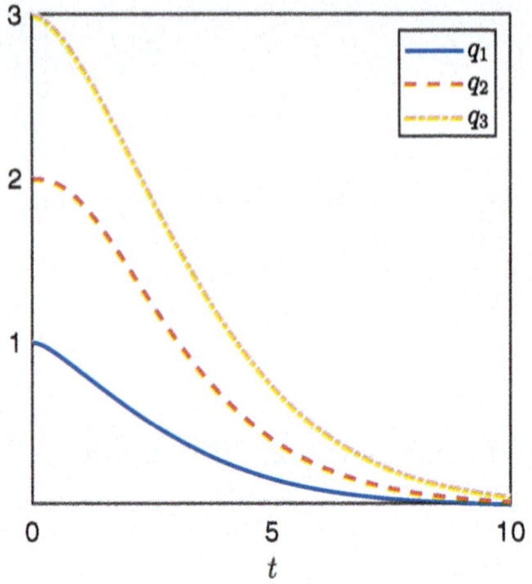

Fig. 1.3 Response of system with coefficients given by Eq. (1.30) and initial conditions $\mathbf{q}(0) = [1 \quad 2 \quad 3]^T$ and $\dot{\mathbf{q}}(0) = \mathbf{0}$. The system is asymptotically stable since all eigenvalues are negative

where the coefficients **c** are determined from the initial conditions. The response is shown in Fig. 1.3.

1.3 Modal Analysis

The equation of motion of damped systems (1.10) comprises the bulk of engineering literature. For this class of systems, the coefficients are symmetric and positive-definite. If **M** is diagonal, the system is said to be inertially decoupled. If **K** is diagonal, the system is elastically decoupled.

These equations are usually coupled and modal analysis is the typical method for decoupling such systems in real space by means of a linear coordinate transformation [2–16].

1.3.1 Undamped Systems

An undamped system is governed by

$$\mathbf{M}\ddot{\mathbf{q}} + \mathbf{K}\mathbf{q} = \mathbf{f}(t). \tag{1.37}$$

The quadratic eigenvalue problem (1.21) associated with this system is

$$(\lambda^2 \mathbf{M} + \mathbf{K})\mathbf{v} = \mathbf{0}. \tag{1.38}$$

Pre-multiplying the above expression by the conjugate transpose $\mathbf{v}^*$ of the eigenvector **v** yields

$$\lambda^2 = -\frac{\mathbf{v}^*\mathbf{M}\mathbf{v}}{\mathbf{v}^*\mathbf{K}\mathbf{v}} < 0 \tag{1.39}$$

due to the positive-definiteness of the coefficient matrices. A logical conclusion is $\lambda = \pm i\omega$, with $\omega > 0$. In addition, each eigenvector can be chosen real and is commonly written as $\mathbf{v} = \mathbf{u} \in \mathbb{R}^n$. The parameter $\omega > 0$ is called a natural frequency, and the corresponding real eigenvector $\mathbf{u} \neq \mathbf{0}$ is termed a mode shape. The quadratic eigenvalue problem is thus expressed as

$$\mathbf{K}\mathbf{u} = \omega^2\mathbf{M}\mathbf{u}. \tag{1.40}$$

The polynomial equation:

$$\det(\mathbf{K} - \omega^2\mathbf{M}) = 0, \tag{1.41}$$

is of degree n in ω^2 so there are n natural frequencies and all mode shapes are real. By convention, the natural frequencies are arranged in the order of increasing magnitude such that $\omega_1 \leq \omega_2 \leq \ldots \leq \omega_n$. The lowest frequency ω_1 is called the fundamental frequency. It is assumed herein that the frequencies are distinct, but the reader is referred to [26] for the case of repeated eigenvalues.

The mode shapes are usually normalized according to

$$\mathbf{u}_l^T\mathbf{M}\mathbf{u}_j = \delta_{lj}, \tag{1.42}$$

where $\delta_{lj} = 0$ if $l \neq j$ or $\delta_{lj} = 1$ if $l = j$. In this case, it is said the mode shapes are mass normalized. It follows that

$$\mathbf{u}_l^T\mathbf{K}\mathbf{u}_j = \omega_l^2\delta_{lj}. \tag{1.43}$$

Define the modal and the (diagonal) spectral matrices, respectively, by

$$\mathbf{U} = \begin{bmatrix}\mathbf{u}_1 & \mathbf{u}_2 & \cdots & \mathbf{u}_n\end{bmatrix}, \ \mathbf{\Omega} = \begin{bmatrix}\omega_1^2 & & & \\ & \omega_2^2 & & \\ & & \ddots & \\ & & & \omega_n^2\end{bmatrix}. \tag{1.44}$$

The generalized orthogonality of the modes can be expressed in a compact form:

$$\mathbf{U}^T\mathbf{M}\mathbf{U} = \mathbf{I}, \ \mathbf{U}^T\mathbf{K}\mathbf{U} = \mathbf{\Omega}. \tag{1.45}$$

The eigenvectors of $\mathbf{K}\mathbf{u} = \omega^2\mathbf{M}\mathbf{u}$ are orthogonal with respect to $\mathbf{M}$ and $\mathbf{K}$. In general, the eigenvectors are not orthogonal in the ordinary sense. Since the eigenvectors of $\mathbf{K}\mathbf{u} = \omega^2\mathbf{M}\mathbf{u}$ are orthogonal with respect to $\mathbf{M}$ and $\mathbf{K}$, they are also linearly independent. Any set of n linearly independent vectors constitutes a basis in n-dimensional Euclidean space and a theorem follows:

Theorem 1.1 *Any real n-dimensional vector can be expressed as a linear combination of the real eigenvectors of* $\mathbf{K}\mathbf{u} = \omega^2\mathbf{M}\mathbf{u}$.

As a consequence, the response $\mathbf{q}(t)$ of an undamped system (1.37) can be written as

$$\mathbf{q}(t) = \sum_{j=1}^{n} p_j(t)\mathbf{u}_j = \begin{bmatrix} \mathbf{u}_1 & \mathbf{u}_2 & \cdots & \mathbf{u}_n \end{bmatrix} \begin{bmatrix} p_1 \\ p_2 \\ \vdots \\ p_n \end{bmatrix} = \mathbf{U}\mathbf{p}(t), \tag{1.46}$$

where the coefficients $p_j(t)$ are termed principal or modal coordinates. The above expression is referred to as modal expansion, mode superposition, or eigenvector expansion of the system response $\mathbf{q}(t)$. This can be regarded as coordinate transformation from $\mathbf{q}$ to $\mathbf{p}$. In terms of $\mathbf{p}$, the undamped equation of motion (1.37) is decoupled:

$$\ddot{\mathbf{p}} + \boldsymbol{\Omega}\mathbf{p} = \mathbf{g}(t). \tag{1.47}$$

The initial values are transformed in accordance with $\mathbf{p}(0) = \mathbf{U}^T\mathbf{M}\mathbf{q}(0)$ and $\dot{\mathbf{p}}(0) = \mathbf{U}^T\mathbf{M}\dot{\mathbf{q}}(0)$. The excitation $\mathbf{f}(t)$ is transformed into

$$\mathbf{g}(t) = \mathbf{U}^T\mathbf{f}(t) = \begin{bmatrix} \mathbf{u}_1^T\mathbf{f}(t) \\ \mathbf{u}_2^T\mathbf{f}(t) \\ \vdots \\ \mathbf{u}_n^T\mathbf{f}(t) \end{bmatrix}. \tag{1.48}$$

The modal equation associated with p_j is

$$\ddot{p}_j + \omega_j^2 p_j = \mathbf{u}_j^T\mathbf{f}(t), \quad j = 1, 2, \ldots, n. \tag{1.49}$$

Each decoupled modal equation can be solved independently, and solution of the original equation is readily recovered from $\mathbf{q} = \mathbf{U}\mathbf{p}$. This decoupling procedure is the time-honored process of modal analysis. Therefore, it follows that

Theorem 1.2 *An undamped system can always be decoupled in real space by modal analysis.*

1.3.1.1 Example 1.3

Consider the undamped system:

$$\ddot{\mathbf{q}} + \begin{bmatrix} 2 & -1 \\ -1 & 2 \end{bmatrix} \mathbf{q} = \mathbf{0}. \tag{1.50}$$

The associated mass-normalized modal matrix is

$$\mathbf{U} = \frac{1}{\sqrt{2}} \begin{bmatrix} 1 & 1 \\ 1 & -1 \end{bmatrix}. \tag{1.51}$$

The system is diagonalized into

$$\ddot{\mathbf{p}} + \begin{bmatrix} 1 & 0 \\ 0 & 3 \end{bmatrix} \mathbf{p} = \mathbf{0}. \tag{1.52}$$

For the initial conditions $\mathbf{q}(0) = [1 \quad 2]^T$ and $\dot{\mathbf{q}}(0) = \mathbf{0}$:

$$\mathbf{q}(t) = \frac{3}{2}\cos(t) \begin{bmatrix} 1 \\ 1 \end{bmatrix} - \frac{1}{2}\cos(\sqrt{3}t) \begin{bmatrix} 1 \\ -1 \end{bmatrix}, \tag{1.53}$$

and it is shown in Fig. 1.4. In modal analysis, a coupled system is decoupled into real and independent sub-systems by a real and invertible coordinate transformation in the configuration space. The coupled and decoupled systems are equivalent. Specifically, a coupled system is converted into independent single-degree-of-freedom oscillators. This process is illustrated in Fig. 1.5 regarding Example 1.3.

1.3.2 Damped Systems

Any undamped system can be decoupled by modal analysis. One might ask if modal analysis is applicable to a damped system (1.10). If the modal transformation $\mathbf{q} = \mathbf{Up}$ is applied to a damped system, Eq. (1.12) is obtained for which the mass and stiffness matrices are diagonalized in accordance with Eq. (1.45). The system is decoupled if and only if the modal damping matrix $\mathbf{D} = \mathbf{U}^T\mathbf{CU}$ is diagonal.

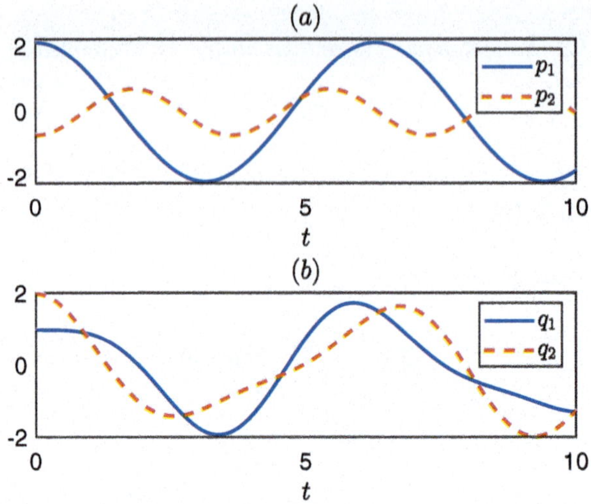

Fig. 1.4 (**a**) Response of decoupled system (1.52) and (**b**) response of undamped system (1.50) with the initial conditions $\mathbf{q}(0) = [1 \quad 2]^T$ and $\dot{\mathbf{q}}(0) = \mathbf{0}$

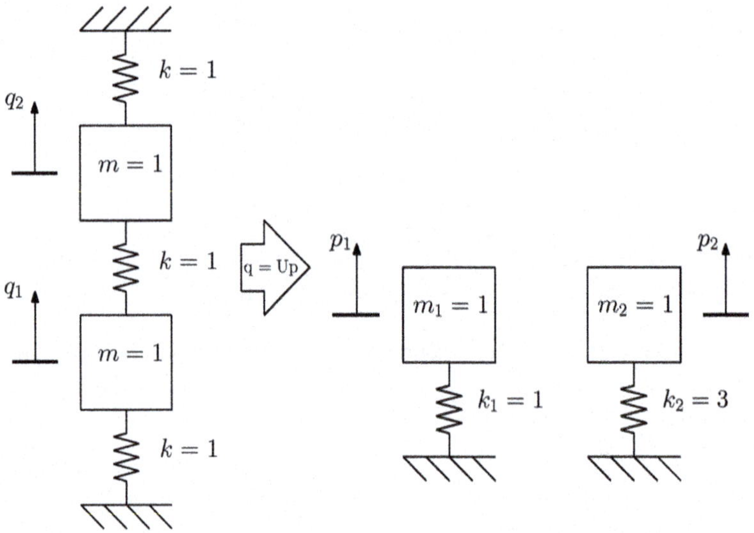

Fig. 1.5 Modal analysis converts the coupled oscillator (1.50) into independent oscillators in (1.52)

In Section 97 of *The Theory of Sound* [1], Lord Rayleigh asserted that a damped system can be decoupled by modal analysis if

$$\mathbf{C} = \eta_0 \mathbf{M} + \eta_1 \mathbf{K} \tag{1.54}$$

for some scalar constants η_0 and η_1. The modal damping matrix $\mathbf{D}$ is diagonal because

$$\mathbf{D} = \mathbf{U}^T \mathbf{C} \mathbf{U} = \mathbf{U}^T (\eta_0 \mathbf{M} + \eta_1 \mathbf{K}) \mathbf{U} = \eta_0 \mathbf{I} + \eta_1 \boldsymbol{\Omega}. \tag{1.55}$$

This requirement, referred to as proportional damping, is sufficient but not necessary for decoupling by modal analysis.

The question remains what subclass of damped systems can be diagonalized by modal analysis. This question was addressed in the seminal paper by Caughey and O'Kelly [17]. A system is termed *classically damped* if and only if the modal damping matrix $\mathbf{D} = \mathbf{U}^T \mathbf{C} \mathbf{U} = [d_{lj}]$ is diagonal. Caughey and O'Kelly analyzed the requirement:

$$\mathbf{C} \mathbf{M}^{-1} \mathbf{K} = \mathbf{K} \mathbf{M}^{-1} \mathbf{C}. \tag{1.56}$$

Orthogonality relations (1.45) imply that $\mathbf{M}^{-1} = \mathbf{U} \mathbf{U}^T$ and $\mathbf{K} = \mathbf{U}^{-T} \boldsymbol{\Omega} \mathbf{U}^{-1}$ so condition (1.56) can be equivalently expressed as

$$\begin{aligned} \mathbf{C} \mathbf{M}^{-1} \mathbf{K} = \mathbf{K} \mathbf{M}^{-1} \mathbf{C} &\Longleftrightarrow \mathbf{C} \mathbf{U} \mathbf{U}^T \mathbf{U}^{-T} \boldsymbol{\Omega} \mathbf{U}^{-1} = \mathbf{U}^{-T} \boldsymbol{\Omega} \mathbf{U}^{-1} \mathbf{U} \mathbf{U}^T \mathbf{C} \\ &\Longleftrightarrow \mathbf{U}^T \mathbf{C} \mathbf{U} \boldsymbol{\Omega} = \boldsymbol{\Omega} \mathbf{U}^T \mathbf{C} \mathbf{U} \\ &\Longleftrightarrow \mathbf{D} \boldsymbol{\Omega} = \boldsymbol{\Omega} \mathbf{D} \\ &\Longleftrightarrow d_{lj} \omega_j^2 = \omega_l^2 d_{lj}. \quad (l, j = 1, 2, \ldots, n) \end{aligned} \tag{1.57}$$

The above equation implies that $d_{lj} = 0$ if $l \neq j$ with the assumption that natural frequencies are distinct. Condition (1.56) is also valid for systems with repeated eigenvalues. This case is treated by orthogonalization in [26]. Hence, $\mathbf{D}$ is diagonal if and only if $\mathbf{K} \mathbf{M}^{-1} \mathbf{C} = \mathbf{C} \mathbf{M}^{-1} \mathbf{K}$. Eq. (1.56) is a necessary and sufficient condition for decoupling a damped system by modal analysis, a process depicted in Fig. 1.6.

Theorem 1.3 *A damped system is decoupled by modal analysis if and only if* $\mathbf{K} \mathbf{M}^{-1} \mathbf{C} = \mathbf{C} \mathbf{M}^{-1} \mathbf{K}$.

It should be mentioned that condition (1.56) can be cast into several equivalent forms, some of which use the inverses of $\mathbf{C}$ or $\mathbf{K}$ [64]. For example, the coefficient matrices can be simultaneously diagonalized by the same matrix $\mathbf{U}$ if and only if

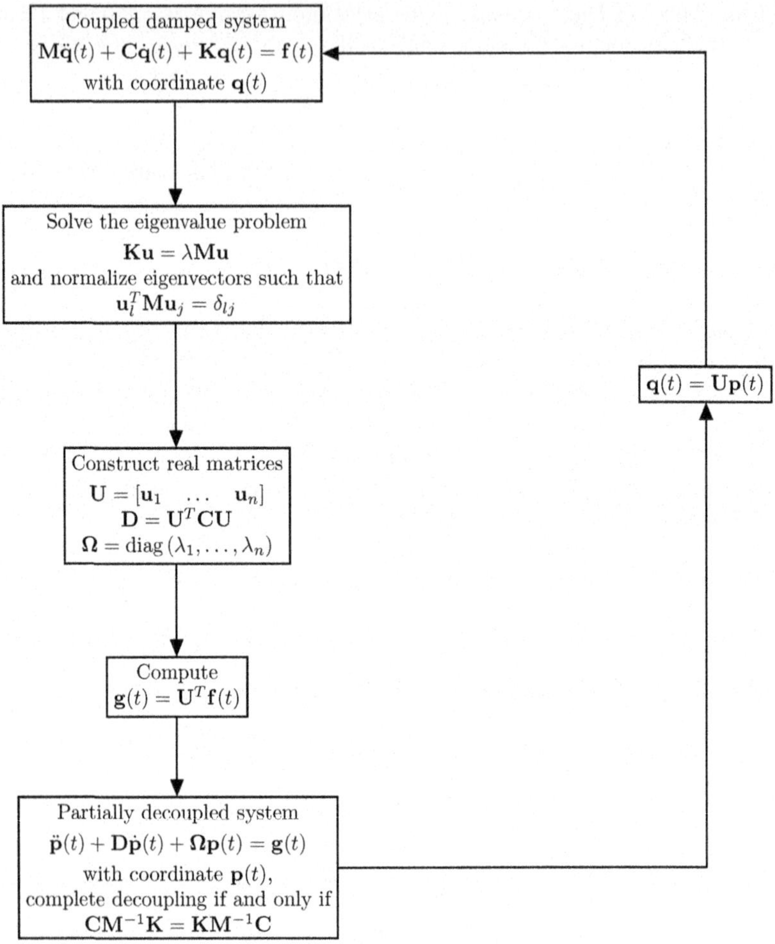

Fig. 1.6 Algorithm for decoupling a damped system by modal analysis

$$\mathbf{M}\mathbf{K}^{-1}\mathbf{C} = \mathbf{C}\mathbf{K}^{-1}\mathbf{M} \tag{1.58}$$

or

$$\mathbf{M}\mathbf{C}^{-1}\mathbf{K} = \mathbf{K}\mathbf{C}^{-1}\mathbf{M}. \tag{1.59}$$

This means all the three system matrices can be treated on equal basis and therefore can be interchanged [64]. Since **M** is assumed non-singular, condition (1.56) is preferred.

1.3.3 Inadequacy of Modal Analysis

In the presence of damping, condition (1.56) implies a linear system may not be decoupled by modal analysis. A system is termed non-classically damped if it cannot be decoupled by modal analysis. Practically speaking, classical damping, i.e., condition (1.56), means that energy dissipation is almost uniformly distributed throughout the system [27, 28, 65].

There is no reason why damping should be classical. This assumption is violated for systems consisting of two or more parts with significantly different levels of damping. Examples of such systems include soil-structure systems [66], base-isolated structures [67–69], and systems in which coupled vibrations of structures and fluids occur. Increasing use of special energy-dissipating devices in control constitutes another important example. In fact, experimental modal testing suggests that no real physical system is strictly classically damped [70]. Thus linear systems are generally non-classical and cannot be decoupled by modal transformation.

1.3.3.1 Example 1.4

Consider a damped system with coefficients

$$\mathbf{M}=\mathbf{I},\ \mathbf{C}=\frac{1}{10}\begin{bmatrix}7 & -1\\ -1 & 2\end{bmatrix},\ \mathbf{K}=\begin{bmatrix}2 & -1\\ -1 & 2\end{bmatrix}. \tag{1.60}$$

While the modal matrix

$$\mathbf{U}=\frac{1}{\sqrt{2}}\begin{bmatrix}1 & 1\\ 1 & -1\end{bmatrix} \tag{1.61}$$

diagonalizes both **M** and **K**, the modal damping matrix is

$$\mathbf{D}=\mathbf{U}^T\mathbf{C}\mathbf{U}=\frac{1}{20}\begin{bmatrix}7 & 5\\ 5 & 11\end{bmatrix}. \tag{1.62}$$

This system is partially decoupled, with coupling ultimately occurring through **D**.

1.3.3.2 Example 1.5

A damped system with

$$\mathbf{M}=\mathbf{I},\ \mathbf{C}=\frac{1}{4}\begin{bmatrix}4 & -1\\ -1 & 4\end{bmatrix},\ \mathbf{K}=\begin{bmatrix}2 & -1\\ -1 & 2\end{bmatrix} \tag{1.63}$$

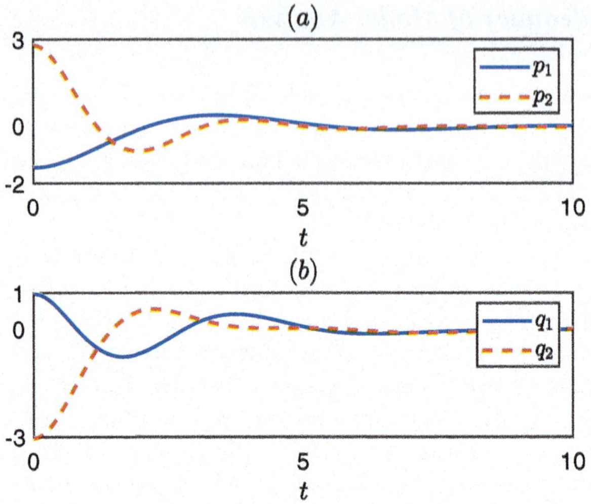

Fig. 1.7 System (1.63) is classically damped, and the homogeneous response can be computed through modal superposition. (**a**) The response of the decoupled system (1.65) and (**b**) the response of the coupled system (1.63) for the initial conditions $\mathbf{q}(0) = [1 \quad -3]^T$ and $\dot{\mathbf{q}}(0) = \mathbf{0}$ are shown. Damping results in exponentially decaying solution

is classically damped because $\mathbf{CK} = \mathbf{KC}$. This means it can be decoupled by the modal transformation $\mathbf{q} = \mathbf{Up}$, where

$$\mathbf{U} = \frac{1}{\sqrt{2}}\begin{bmatrix} 1 & 1 \\ 1 & -1 \end{bmatrix}. \tag{1.64}$$

The decoupled system is

$$\ddot{\mathbf{p}} + \frac{1}{4}\begin{bmatrix} 3 & 0 \\ 0 & 5 \end{bmatrix}\dot{\mathbf{p}} + \begin{bmatrix} 1 & 0 \\ 0 & 3 \end{bmatrix}\mathbf{p} = \mathbf{U}^T\mathbf{f}(t). \tag{1.65}$$

A homogeneous response of this system is shown in Fig. 1.7.

1.3.4 Proportional and Caughey Damping

Now, another point of view may be brought up regarding proportional damping. If $\mathbf{C} = \eta_0\mathbf{M} + \eta_1\mathbf{K}$, then

$$\begin{aligned} \mathbf{CM}^{-1}\mathbf{K} &= (\eta_0\mathbf{M} + \eta_1\mathbf{K})\mathbf{M}^{-1}\mathbf{K} = \eta_0\mathbf{K} + \eta_1\mathbf{KM}^{-1}\mathbf{K} \\ &= \mathbf{KM}^{-1}(\eta_0\mathbf{M} + \eta_1\mathbf{K}) = \mathbf{KM}^{-1}\mathbf{C}. \end{aligned} \tag{1.66}$$

Hence, a system with proportional damping $\mathbf{C} = \eta_0\mathbf{M} + \eta_1\mathbf{K}$ is classically damped. An extension of proportional damping (1.54) involves a power series expansion [17]:

$$\mathbf{C} = \mathbf{M}\sum_{j=0}^{n-1} \eta_j (\mathbf{M}^{-1}\mathbf{K})^j. \tag{1.67}$$

Of course, the power series need not contain powers higher than $n - 1$ because of the Cayley–Hamilton theorem [71]. A system with such a damping matrix, termed Caughey damping [4], is also classically damped because

$$\begin{aligned}
\mathbf{C}\mathbf{M}^{-1}\mathbf{K} &= \left\{\mathbf{M}\sum_{j=0}^{n-1} \eta_j (\mathbf{M}^{-1}\mathbf{K})^j\right\}\mathbf{M}^{-1}\mathbf{K} = \mathbf{M}\sum_{j=0}^{n-1} \eta_j (\mathbf{M}^{-1}\mathbf{K})^{j+1} \\
&= \mathbf{M}(\mathbf{M}^{-1}\mathbf{K})\sum_{j=0}^{n-1} \eta_j (\mathbf{M}^{-1}\mathbf{K})^j = \mathbf{K}\mathbf{M}^{-1}\mathbf{M}\sum_{j=0}^{n-1} \eta_j (\mathbf{M}^{-1}\mathbf{K})^j \\
&= \mathbf{K}\mathbf{M}^{-1}\mathbf{C}.
\end{aligned} \tag{1.68}$$

This is an extension of proportional damping because, if only the first two terms of the power series are retained,

$$\mathbf{C} = \mathbf{M}\eta_0(\mathbf{M}^{-1}\mathbf{K})^0 + \mathbf{M}\eta_1(\mathbf{M}^{-1}\mathbf{K})^1 = \eta_0\mathbf{M} + \eta_1\mathbf{K}, \tag{1.69}$$

which reduces to proportional damping.

Caughey and O'Kelly [17] showed that, when the natural frequencies are distinct, a system is classically damped if and only if $\mathbf{C}$ can be expressed as the power series in (1.67). This may be easily seen by expressing (1.67) in an equivalent form. Use the orthogonality relations (1.45) to rewrite (1.67) as

$$\mathbf{C} = \mathbf{U}^{-T}\mathbf{U}^{-1}\sum_{j=0}^{n-1} \eta_j (\mathbf{U}\mathbf{U}^T\mathbf{K})^j. \tag{1.70}$$

Upon substitution into $\mathbf{D} = \mathbf{U}^T\mathbf{C}\mathbf{U}$ and rearrangement, the modal damping matrix may be expressed as a power series in $\boldsymbol{\Omega}$:

$$\mathbf{D} = \sum_{j=0}^{n-1} \eta_j \boldsymbol{\Omega}^j. \tag{1.71}$$

Remark 1.1 Expressing $\mathbf{C}$ as the power series in (1.67) is equivalent to expressing $\mathbf{D}$ as the power series in (1.71).

Therefore, when the natural frequencies are distinct, a system is classically damped if and only if $\mathbf{C}$ can be expressed as power series in $\mathbf{M}^{-1}\mathbf{K}$ or, equivalently, $\mathbf{D}$ is a power series in $\boldsymbol{\Omega}$. To prove this, observe that a system is classically damped if and only if the modal damping matrix $\mathbf{D} = \mathbf{U}^T\mathbf{C}\mathbf{U}$ is diagonal. Suppose $\mathbf{D}$ is represented by the power series in (1.71). Since $\boldsymbol{\Omega}$ is diagonal, $\mathbf{D}$ is also diagonal and the system is classically damped.

On the contrary, assume that the system is classically damped. Then $\mathbf{D}$ is diagonal and it remains to show that it can be represented by (1.71). Note that (1.71) means that

$$\begin{bmatrix} d_{11} \\ d_{22} \\ \vdots \\ d_{nn} \end{bmatrix} = \begin{bmatrix} 1 & \omega_1^2 & \cdots & \omega_1^{2n-2} \\ 1 & \omega_2^2 & \cdots & \omega_2^{2n-2} \\ \vdots & \vdots & \ddots & \vdots \\ 1 & \omega_n^2 & \cdots & \omega_n^{2n-2} \end{bmatrix} \begin{bmatrix} \eta_0 \\ \eta_1 \\ \vdots \\ \eta_{n-1} \end{bmatrix} = \mathbf{V}_A \begin{bmatrix} \eta_0 \\ \eta_1 \\ \vdots \\ \eta_{n-1} \end{bmatrix}, \tag{1.72}$$

where $\mathbf{V}_A$ is the Vandermonde matrix, for which [71]

$$\det(\mathbf{V}_A) = (\omega_2^2-\omega_1^2)(\omega_3^2-\omega_1^2)(\omega_3^2-\omega_2^2)\ldots(\omega_n^2-\omega_1^2) = \prod_{\substack{l,j=1 \\ l>j}}^{2n} (\omega_l^2-\omega_j^2). \tag{1.73}$$

The Vondermonde determinant $\det(\mathbf{V}_A) \neq 0$ if and only if the frequencies are distinct. With the assumptions of distinct frequencies, the coefficients in the power series (1.67) can be determined uniquely from

$$\begin{bmatrix} \eta_0 \\ \eta_1 \\ \vdots \\ \eta_{n-1} \end{bmatrix} = \mathbf{V}_A^{-1} \begin{bmatrix} d_{11} \\ d_{22} \\ \vdots \\ d_{nn} \end{bmatrix}. \tag{1.74}$$

Theorem 1.4 *If the natural frequencies are distinct, a system is classically damped if and only if* $\mathbf{C}$ *is a power series in* $\mathbf{M}^{-1}\mathbf{K}$ *or, equivalently, the modal damping matrix* $\mathbf{D}$ *is a power series in* $\boldsymbol{\Omega}$.

A classically damped system whose damping matrix **C** is not a power series in $\mathbf{M}^{-1}\mathbf{K}$ should have repeated frequencies. Therefore,

Remark 1.2 Proportional damping or Caughey damping is not necessary for a system to be classically damped.

1.3.4.1 Example 1.6

Consider the classically damped system from Example 1.5. The natural frequencies are distinct and equal to $\omega_1 = 1$ and $\omega_2 = \sqrt{3}$. This means that it is possible to write the damping matrix as $\mathbf{C} = \eta_0\mathbf{M} + \eta_1\mathbf{K}$. To the determine these coefficients, compute the associated Vandermonde matrix:

$$\mathbf{V}_A = \begin{bmatrix} 1 & 1 \\ 1 & 3 \end{bmatrix}. \tag{1.75}$$

It can be checked that $\det(\mathbf{V}_A) = 2 = \omega_2^2 - \omega_1^2$. The coefficients η_0 and η_1 can then be determined from

$$\begin{bmatrix} \eta_0 \\ \eta_1 \end{bmatrix} = \mathbf{V}_A^{-1} \begin{bmatrix} 3/4 \\ 5/4 \end{bmatrix} = \begin{bmatrix} 1/2 \\ 1/4 \end{bmatrix}. \tag{1.76}$$

It is also possible to check that $\mathbf{D} = \eta_0\mathbf{I} + \eta_1\boldsymbol{\Omega}$.

1.3.4.2 Example 1.7

Consider the damped system:

$$\mathbf{M} = \mathbf{I},\ \mathbf{C} = \frac{1}{3}\begin{bmatrix} 22 & 1 & 7 \\ 1 & 28 & 1 \\ 7 & 1 & 22 \end{bmatrix},\ \mathbf{K} = \frac{1}{6}\begin{bmatrix} 7 & -2 & 1 \\ -2 & 10 & -2 \\ 1 & -2 & 7 \end{bmatrix}. \tag{1.77}$$

The mass-normalized modal matrix is

$$\mathbf{U} = \begin{bmatrix} 1/\sqrt{3} & 1/\sqrt{2} & 1/\sqrt{6} \\ 1/\sqrt{3} & 0 & -2/\sqrt{6} \\ 1/\sqrt{3} & -1/\sqrt{2} & 1/\sqrt{6} \end{bmatrix} \tag{1.78}$$

and the spectral and modal damping matrices are

$$\boldsymbol{\Omega} = \begin{bmatrix} 1 & 0 & 0 \\ 0 & 1 & 0 \\ 0 & 0 & 2 \end{bmatrix}, \quad \mathbf{D} = \begin{bmatrix} 10 & 0 & 0 \\ 0 & 5 & 0 \\ 0 & 0 & 9 \end{bmatrix}, \tag{1.79}$$

respectively. This system is classically damped, but it has repeated natural frequencies. Therefore, $\mathbf{C}$ cannot be expressed as the power series in (1.67). Indeed, the linear system of equations $\mathbf{C} = \eta_0\mathbf{I} + \eta_1\mathbf{K} + \eta_2\mathbf{K}^2$ is indeterminate.

1.3.5 Remarks on Hamiltonian Approach

Since modal analysis is not applicable in general, it is always possible, in theory, to investigate a non-classically damped system with the Hamiltonian approach, whereby the second-order Eq. (1.11) is recast into a first-order system of dimension $2n$. The state-space representation is one such approach.

As clarified by Foss and others [18–20], modal analysis can be extended to a process of complex modal analysis in state space to treat non-classically damped systems. The equation of motion of an n-degree-of-freedom damped system (1.10) can be cast in $2n$-dimensional state space as a first-order equation:

$$\dot{\mathbf{x}} = \begin{bmatrix} \dot{\mathbf{q}} \\ \ddot{\mathbf{q}} \end{bmatrix} = \begin{bmatrix} \mathbf{0} & \mathbf{I} \\ -\mathbf{M}^{-1}\mathbf{K} & -\mathbf{M}^{-1}\mathbf{C} \end{bmatrix} \begin{bmatrix} \mathbf{q} \\ \dot{\mathbf{q}} \end{bmatrix} + \begin{bmatrix} \mathbf{0} \\ \mathbf{M}^{-1}\mathbf{f}(t) \end{bmatrix} = \mathbf{A}\mathbf{x} + \mathbf{b}(t). \tag{1.80}$$

Some state equations may be constructed to have symmetric but not positive-definite coefficients. For example, a damped system may be represented in state space by an equation with symmetric coefficients as

$$\begin{bmatrix} \mathbf{C} & \mathbf{M} \\ \mathbf{M} & \mathbf{0} \end{bmatrix} \begin{bmatrix} \dot{\mathbf{q}} \\ \ddot{\mathbf{q}} \end{bmatrix} = \begin{bmatrix} \mathbf{K} & \mathbf{0} \\ \mathbf{0} & -\mathbf{M} \end{bmatrix} \begin{bmatrix} \mathbf{q} \\ \dot{\mathbf{q}} \end{bmatrix} + \begin{bmatrix} \mathbf{f}(t) \\ \mathbf{0} \end{bmatrix} \tag{1.81}$$

while

$$\begin{bmatrix} \mathbf{0} \\ \mathbf{1} \end{bmatrix}^T \begin{bmatrix} \mathbf{K} & \mathbf{0} \\ \mathbf{0} & -\mathbf{M} \end{bmatrix} \begin{bmatrix} \mathbf{0} \\ \mathbf{1} \end{bmatrix} = -\mathbf{1}^T\mathbf{M}\mathbf{1} < 0, \tag{1.82}$$

where $\mathbf{1}$ is an n-dimensional vector with all entries equal to one. The lack of definiteness means that the eigenvalues are not necessarily real. Other state-space representations may be constructed, but they all share the same eigenvalues and thus are considered equivalent.

Complex modal analysis is a method for decoupling the state equation (1.80) by means of a state transformation. The state-space approach has never appealed to practicing engineers. There are several reasons for this situation. A common excuse is that it is computationally more involved because the dimension of state space is twice the number of degrees of freedom. Another reason is that complex modal analysis may not even decouple the state equation of a classically damped system (see Example 1.8).

More importantly, if decoupling is achieved by complex modal analysis, the resulting independent equations are generally complex. There is serious absence of physical insight associated with different elements of complex modal analysis, while classical modal analysis is amenable to physical interpretation. For example, a modal vector represents the profile of vibration. Even the symmetric eigenvalue problem (1.40) can be interpreted as the problem of finding the principal axes of an n-dimensional ellipsoid [2].

1.3.5.1 Example 1.8

Consider a damped system (1.10) with coefficients

$$\mathbf{M} = \mathbf{I},\ \mathbf{C} = \begin{bmatrix} 3 & -1 \\ -1 & 3 \end{bmatrix},\ \mathbf{K} = \frac{1}{2}\begin{bmatrix} 5 & -3 \\ -3 & 5 \end{bmatrix}. \tag{1.83}$$

Such system can be decoupled by modal analysis into

$$\ddot{\mathbf{p}} + \begin{bmatrix} 2 & 0 \\ 0 & 4 \end{bmatrix}\dot{\mathbf{p}} + \begin{bmatrix} 1 & 0 \\ 0 & 4 \end{bmatrix}\mathbf{p} = \mathbf{g}(t). \tag{1.84}$$

The state companion matrix is

$$\mathbf{A} = \begin{bmatrix} \mathbf{0} & \mathbf{I} \\ -\mathbf{M}^{-1}\mathbf{K} & -\mathbf{M}^{-1}\mathbf{C} \end{bmatrix} = \begin{bmatrix} 0 & 0 & 1 & 0 \\ 0 & 0 & 0 & 1 \\ -5/2 & 3/2 & -3 & 1 \\ 3/2 & -5/2 & 1 & -3 \end{bmatrix}. \tag{1.85}$$

The state companion matrix $\mathbf{A}$ is diagonalizable if and only if it has four linearly independent eigenvectors (matrix is non-defective) [71]. The eigenvalues of $\mathbf{A}$ are $\lambda_1 = \lambda_2 = -1$ and $\lambda_3 = \lambda_4 = -2$. However, there is only one eigenvector associated with λ_1 and λ_2. The same is true for λ_3 and λ_4. Therefore, $\mathbf{A}$ is defective and cannot be diagonalized. As a result, the state equation of this system cannot be decoupled by complex modal analysis. The state companion matrix must be non-defective in order for complex modal analysis to achieve complete decoupling.

1.4 Decoupling by Similarity Transformation

Modal analysis is the time-honored technique for decoupling classically damped systems (1.10). It consists of a non-singular linear coordinate transformation that decouples a system in real configuration space. Even though it is not applicable to linear non-conservative systems (1.11), it would be preferable if a non-singular linear coordinate transformation could still be used to decouple general linear systems in real configuration space.

Mathematically, decoupling a damped system by modal analysis is equivalent to the problem of simultaneously diagonalizing three matrices by a congruence transformation. Diagonalizing a square matrix $\mathbf{B}$ by congruence transformation is attained by a non-singular transformation matrix $\mathbf{V}$ such that $\mathbf{V}^T\mathbf{B}\mathbf{V}$ is diagonal. However, congruence transformations cannot be used to advantage on non-conservative systems, whose coefficient matrices lack symmetry.

The most common technique in linear algebra for diagonalizing a matrix is through similarity transformation [71–73]. If there exists a non-singular matrix $\mathbf{V}$ such that $\mathbf{V}^{-1}\mathbf{B}\mathbf{V}$ is diagonal, $\mathbf{B}$ is said to be diagonalizable by similarity transformation. If $\mathbf{B}$ has distinct eigenvalues, it is diagonalizable. More generally, $\mathbf{B}$ is diagonalizable if and only if it has a full set of independent eigenvectors.

The application of simultaneous diagonalization by similarity transformations was summarized by Caughey and Ma [24]. Two diagonalizable matrices $\mathbf{B}$, $\mathbf{E}$ are simultaneously diagonalizable by a common similarity transformation if and only if there exists a non-singular matrix $\mathbf{V}$ such that $\mathbf{V}^{-1}\mathbf{B}\mathbf{V}$ and $\mathbf{V}^{-1}\mathbf{E}\mathbf{V}$ are both diagonal. The problem of simultaneous diagonalization by similarity transformation is addressed in [71], and the main result is

Theorem 1.5 *Two diagonalizable matrices are simultaneously diagonalizable by a similarity transformation if and only if they commute in multiplication.*

Theorem 1.5 implies that

Corollary 1.1 *A family of matrices can be simultaneously diagonalized by a common similarity transformation if and only if each of its members is diagonalizable and any two members commute in multiplication.*

The coefficient matrices of a second-order linear system (1.11) can be simultaneously diagonalized by a common similarity transformation if and only if $\mathbf{A}_2$, $\mathbf{A}_1$, $\mathbf{A}_0$ are diagonalizable and pairwise commutative. Since $\mathbf{A}_2$ is non-singular, the same matrices can be diagonalized if and only if $\mathbf{I}$, $\mathbf{A}_2^{-1}\mathbf{A}_1$, and $\mathbf{A}_2^{-1}\mathbf{A}_0$ are simultaneously diagonalizable, which requires that

$$\mathbf{A}_1\mathbf{A}_2^{-1}\mathbf{A}_0 = \mathbf{A}_0\mathbf{A}_2^{-1}\mathbf{A}_1. \tag{1.86}$$

If this condition holds, there is a non-singular matrix $\mathbf{V}$ that simultaneously diagonalizes all coefficients.

Theorem 1.6 *A linear system is decoupled by a common similarity transformation if and only if* $\mathbf{A}_1\mathbf{A}_2^{-1}\mathbf{A}_0 = \mathbf{A}_0\mathbf{A}_2^{-1}\mathbf{A}_1$.

Condition (1.86) contains the result established by Caughey and O'Kelly (1.56) as special case. In the event that a damped system is classical, a real matrix can be chosen to diagonalize the system simultaneously [24].

When Theorem 1.6 is satisfied, the linear coordinate transformation $\mathbf{q} = \mathbf{V}\mathbf{p}$ transforms the system (1.11) into

$$\mathbf{V}^{-1}\mathbf{A}_2\mathbf{V}\ddot{\mathbf{p}}(t) + \mathbf{V}^{-1}\mathbf{A}_1\mathbf{V}\dot{\mathbf{p}}(t) + \mathbf{V}^{-1}\mathbf{A}_0\mathbf{V}\mathbf{p}(t) = \mathbf{V}^{-1}\mathbf{f}(t), \tag{1.87}$$

where the coefficients $\mathbf{V}^{-1}\mathbf{A}_2\mathbf{V}$, $\mathbf{V}^{-1}\mathbf{A}_1\mathbf{V}$, and $\mathbf{V}^{-1}\mathbf{A}_0\mathbf{V}$ are diagonal. Therefore, system (1.11) has been decoupled into (1.87).

The question remains how to find the invertible decoupling matrix $\mathbf{V}$. Because $\mathbf{A}_2$, $\mathbf{A}_1$, and $\mathbf{A}_0$ commute in multiplication, the diagonalizing matrix $\mathbf{V}$ is their common eigenvector matrix [71]. Computation of the eigenvectors of only one of the coefficients is necessary to determine $\mathbf{V}$. An algorithm for decoupling a linear system by similarity transformations is shown in Fig. 1.8.

1.4.1 Example 1.9

Consider a linear system:

$$\ddot{\mathbf{q}} + \begin{bmatrix} -3 & -4 \\ 2 & 3 \end{bmatrix}\dot{\mathbf{q}} + \begin{bmatrix} 0 & -2 \\ 1 & 3 \end{bmatrix}\mathbf{q} = \mathbf{f}(t). \tag{1.88}$$

Each coefficient matrix is diagonalizable and they satisfy $\mathbf{A}_1\mathbf{A}_0 = \mathbf{A}_0\mathbf{A}_1$. Therefore, the coefficients share the same eigenvector matrix:

$$\mathbf{V} = \begin{bmatrix} 1 & 2 \\ -1 & -1 \end{bmatrix}. \tag{1.89}$$

The coordinate transformation $\mathbf{q} = \mathbf{V}\mathbf{p}$ decouples the system into

$$\ddot{\mathbf{p}} + \begin{bmatrix} 1 & 0 \\ 0 & -1 \end{bmatrix} \dot{\mathbf{p}} + \begin{bmatrix} 2 & 0 \\ 0 & 1 \end{bmatrix} \mathbf{p} = \mathbf{V}^{-1}\mathbf{f}(t). \tag{1.90}$$

1.4.2 Example 1.10

Consider a non-conservative linear system (1.11) with coefficients

$$\mathbf{A}_2 = \mathbf{I},\ \mathbf{A}_1 = \begin{bmatrix} 1 & 1 \\ -2 & 3 \end{bmatrix},\ \mathbf{A}_0 = \begin{bmatrix} -7 & 4 \\ -8 & 1 \end{bmatrix} \tag{1.91}$$

and initial conditions

$$\mathbf{q}(0) = \begin{bmatrix} 1 \\ 2 \end{bmatrix},\ \dot{\mathbf{q}}(0) = \begin{bmatrix} -1 \\ 1 \end{bmatrix}. \tag{1.92}$$

These coefficients satisfy Corollary 1.1, so they can be simultaneously diagonalized by a common similarity transformation. The diagonalizing matrix is

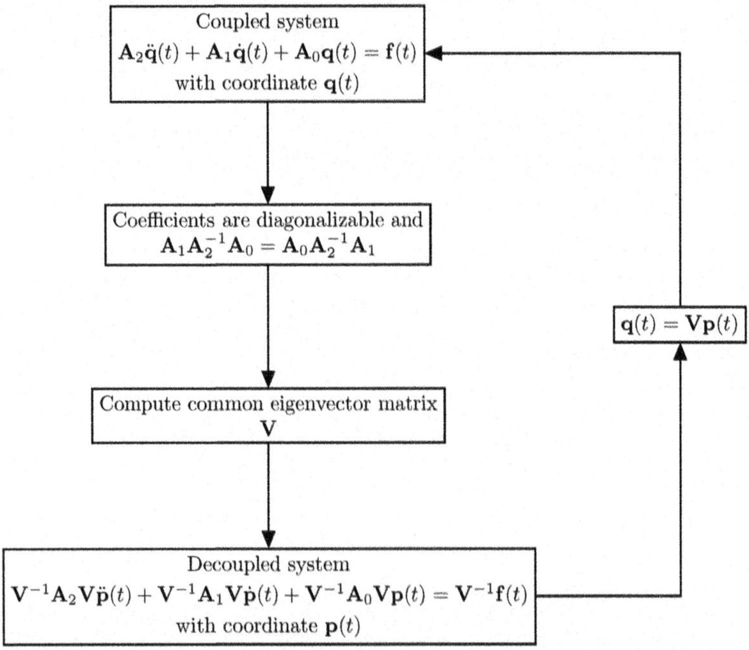

Fig. 1.8 Algorithm for decoupling a linear system by similarity transformations

$$\mathbf{V} = \begin{bmatrix} 1 & 1 \\ 1+i & 1-i \end{bmatrix}. \tag{1.93}$$

To decouple by similarity transformation, define a coordinate transformation by $\mathbf{q} = \mathbf{V}\mathbf{p}$. The decoupled system (1.12) has coefficients

$$\mathbf{D} = \mathbf{V}^{-1}\mathbf{A}_1\mathbf{V} = \begin{bmatrix} 2+i & 0 \\ 0 & 2-i \end{bmatrix}, \ \boldsymbol{\Omega} = \mathbf{V}^{-1}\mathbf{A}_0\mathbf{V}\begin{bmatrix} -3+4i & 0 \\ 0 & -3-4i \end{bmatrix}. \tag{1.94}$$

The initial conditions of the decoupled system are

$$\mathbf{p}(0) = \mathbf{V}^{-1}\mathbf{q}(0) = \frac{1}{2}\begin{bmatrix} 1-i \\ 1+i \end{bmatrix}, \ \dot{\mathbf{p}}(0) = \mathbf{V}^{-1}\dot{\mathbf{q}}(0) = \frac{1}{2}\begin{bmatrix} -1-2i \\ -1+2i \end{bmatrix}. \tag{1.95}$$

The decoupled system and its initial conditions as well as the decoupling transformation are complex. One can still solve the decoupled equations, but physical insight is greatly diminished by such decoupling process.

1.4.3 Symmetrizable Systems

The study of symmetrizability probably represents the first serious attempt in the investigation of non-conservative linear systems. This technique, developed by Huseyin and Leipholz [21] as well as Inman [22], concerns the simultaneous reduction of $\mathbf{A}_2$, $\mathbf{A}_1$, and $\mathbf{A}_0$ to symmetric matrices by a common similarity transformation. A matrix is said to be symmetrizable if and only if it can be expressed as the product of two symmetric matrices one of which is positive-definite. It has been shown by Tausskey and Zassenhaus [74] that every real square matrix can be represented as the product of two real symmetric matrices.

When such procedure is possible, many of the results available for symmetric systems can be applied to non-symmetric systems. For instance, Ahmadian and Inman [23] extended condition (1.56) for symmetrizable non-conservative systems, a special case of (1.86). However, the simultaneous reduction of $\mathbf{A}_2$, $\mathbf{A}_1$, and $\mathbf{A}_0$ to symmetric matrices is not applicable in general, and Inman [22] determined necessary and sufficient conditions for the existence of symmetrizing transformations (e.g., system in Example 1.10 is not symmetrizable [24]). Even after a system has been symmetrized, it generally still cannot be decoupled by modal analysis.

1.4.4 Partial Decoupling: Simultaneous Triangularization

Condition (1.86) imposes a serious restriction on the coefficient matrices so most non-conservative systems cannot be decoupled by common similarity transformations. Even if condition (1.86) is not satisfied, system (1.11) may be partially decoupled when $\mathbf{A}_2$, $\mathbf{A}_1$, and $\mathbf{A}_0$ are simultaneously reduced to upper triangular forms [24, 25]. If the coefficient matrices can be upper triangularized simultaneously, the system is said to be solvable in configuration space because it can be solved by back-substitution afterward.

Any square matrix is similar to an upper triangular matrix. In accordance with Schur decomposition (see Theorem 2.3.1 in [71]), given any square matrix $\mathbf{B}$ there exists a unitary matrix $\mathbf{V}$ such that $\mathbf{V}^{-1}\mathbf{B}\mathbf{V}$ is upper triangular whose diagonal entries are the eigenvalues of $\mathbf{B}$. A matrix $\mathbf{V}$ is unitary if $\mathbf{V}^{-1} = \mathbf{V}^* = \overline{\mathbf{V}}^T$.

The simultaneous triangularization of a family of matrices is based upon the concept of Lie algebra [75]. The notation $\mathbb{L}(\mathbf{A}_0, \mathbf{A}_1, \ldots, \mathbf{A}_m)$ is employed to denote the Lie algebra generated by the matrices $\mathbf{A}_0, \mathbf{A}_1, \ldots, \mathbf{A}_m$. A family of square matrices of order n constitutes a Lie algebra if, for any two matrices in $\mathbb{L}$, the commutator product

$$[\mathbf{A}_0, \mathbf{A}_1] = \mathbf{A}_0\mathbf{A}_1 - \mathbf{A}_1\mathbf{A}_0 \tag{1.96}$$

belongs to $\mathbb{L}$. A Lie algebra $\mathbb{L}$ is associated with its derived sequence, defined inductively as follows:

$$\mathbb{L}^0 = \mathbb{L}, \tag{1.97}$$

$$\mathbb{L}^{j+1} = \{[\mathbf{A}, \mathbf{B}] : \mathbf{A}, \mathbf{B} \in \mathbb{L}^j\}, \ j \geq 0. \tag{1.98}$$

A Lie algebra $\mathbb{L}$ is said to be solvable if there exists an integer j such that $\mathbb{L}^j = \{\mathbf{0}\}$. The class of solvable Lie algebra includes the class of pairwise commuting matrices, for which $\mathbb{L}^1 = \{\mathbf{0}\}$ [24, 25, 75]. An interesting property for solvability is the following statement [76, 77]:

Lemma 1.1 *A matrix Lie algebra $\mathbb{L}$ is solvable if and only if there exists a non-singular matrix $\mathbf{V}$ such that $\mathbf{V}^{-1}\mathbf{B}\mathbf{V}$ is upper triangular for all $\mathbf{B}$ in the Lie algebra.*

The above assertion implies the following theorem:

Theorem 1.7 *A non-conservative system* (1.11) *is solvable if and only if the coefficient matrices $\mathbf{A}_2$, $\mathbf{A}_1$, and $\mathbf{A}_0$ generate a solvable Lie algebra.*

In particular, if $\mathbb{L}^1(\mathbf{A}_2, \mathbf{A}_1, \mathbf{A}_0) = \{\mathbf{0}\}$ and $\mathbf{A}_2$, $\mathbf{A}_1$, and $\mathbf{A}_0$ are diagonalizable, then system (1.11) can be diagonalized [24]. In general, the triangularizing matrix $\mathbf{V}$ is complex. For a classical system, the triangularizing matrix can be chosen to be real.

1.4.4.1 Example 1.11

Consider a two-degree-of-freedom system (1.11) with coefficients

$$\mathbf{A}_2 = \begin{bmatrix} 2 & -1 \\ 1 & 0 \end{bmatrix}, \quad \mathbf{A}_1 = \begin{bmatrix} 1 & 0 \\ 1 & 0 \end{bmatrix}, \quad \mathbf{A}_0 = \begin{bmatrix} 1 & -1 \\ 0 & 0 \end{bmatrix}. \tag{1.99}$$

This system cannot be decoupled by a common similarity transformation, but it may be upper triangularizable. The Lie algebra generated by $\mathbf{A}_2$, $\mathbf{A}_1$, and $\mathbf{A}_0$ is

$$\mathbb{L}(\mathbf{A}_2, \mathbf{A}_1, \mathbf{A}_0) = \{\mathbf{0}, \mathbf{A}_2, \mathbf{A}_1, \mathbf{A}_0, \mathbf{B}, -\mathbf{B}\} \tag{1.100}$$

since

$$[\mathbf{A}_2, \mathbf{A}_1] = [\mathbf{A}_0, \mathbf{A}_2] = [\mathbf{A}_0, \mathbf{A}_1] = \begin{bmatrix} -1 & 1 \\ -1 & 1 \end{bmatrix} = \mathbf{B}, \tag{1.101}$$

$$[\mathbf{A}_1, \mathbf{B}] = [\mathbf{B}, \mathbf{A}_0] = \mathbf{B}, \tag{1.102}$$

$$[\mathbf{A}_2, \mathbf{B}] = \mathbf{0}. \tag{1.103}$$

As a consequence, $\mathbb{L}^0 = \mathbb{L}$, $\mathbb{L}^1 = \{\mathbf{0}, \mathbf{B}, -\mathbf{B}\}$ and $\mathbb{L}^2 = \{\mathbf{0}\}$. Therefore $\mathbb{L}(\mathbf{A}_2, \mathbf{A}_1, \mathbf{A}_0)$ is solvable and $\mathbf{A}_2$, $\mathbf{A}_1$, and $\mathbf{A}_0$ can be converted into upper triangular forms by a common similarity transformation. The triangularizing matrix is [24]

$$\mathbf{V} = \begin{bmatrix} 1 & 1 \\ 1 & 0 \end{bmatrix}. \tag{1.104}$$

The coordinate transformation $\mathbf{q} = \mathbf{V}\mathbf{p}$ transforms the original system into

$$\begin{bmatrix} 1 & 1 \\ 0 & 1 \end{bmatrix} \ddot{\mathbf{p}} + \begin{bmatrix} 1 & 1 \\ 0 & 0 \end{bmatrix} \dot{\mathbf{p}} + \begin{bmatrix} 0 & 0 \\ 0 & 1 \end{bmatrix} \mathbf{p} = \mathbf{V}^{-1}\mathbf{f}(t). \tag{1.105}$$

Such system of differential equations may be solved by backward substitution.

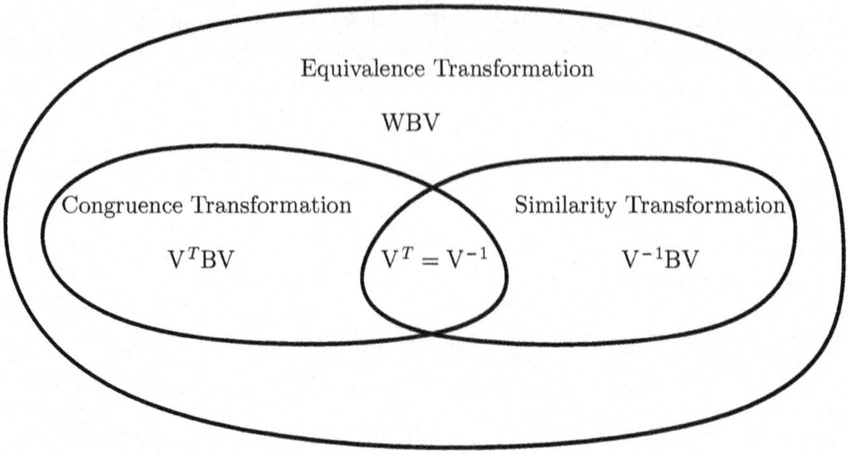

Fig. 1.9 A diagram depicting the hierarchy of non-singular linear configuration-space decoupling techniques. Equivalence transformations are the most general non-singular linear transformation. Similarity transformations belong to a subset of equivalence transformations and are reviewed in Sect. 1.4. Classical modal analysis, reviewed in Sect. 1.3, utilizes a congruence transformation. It is a similarity transformation when the modal matrix is orthogonal: $\mathbf{U}^T = \mathbf{U}^{-1}$

1.5 Decoupling by Equivalence Transformation

An examination of the restrictions involved in the methods in Sects. 1.3 and 1.4 reveals that only a small subclass of linear systems can indeed be treated. In this regard, the most general non-singular linear decoupling transformation one may use is based on equivalence transformations [26].

If a square matrix **B** is diagonalized by an equivalence transformation, there exist two non-singular matrices **V** and **W** such that **WBV** is diagonal. An equivalence transformation preserves the rank of the matrix. If $\mathbf{W} = \mathbf{V}^{-1}$, the equivalence transformation is a similarity transformation, expounded in Sect. 1.4. In the event that $\mathbf{W} = \mathbf{V}^T$, the equivalence transformation is a congruence transformation, of which the classical modal transformation presented in Sect. 1.3 is an example. A congruence transformation is also a similarity transformation if **V** is an orthogonal matrix.

Equivalence transformations that are neither similarity nor congruence transformations are rarely used in structural analysis [26]. Nevertheless, equivalence transformations are the most general non-singular transformation in the hierarchy of linear decoupling techniques, as depicted in Fig. 1.9.

1.5.1 Undamped Non-gyroscopic Systems

The equation of motion of an undamped non-gyroscopic system (for which $\mathbf{A}_1 = \mathbf{0}$) is

$$\mathbf{A}_2\ddot{\mathbf{q}} + \mathbf{A}_0\mathbf{q} = \mathbf{f}(t). \tag{1.106}$$

The above system can be decoupled by an equivalence transformation if and only if there exist two non-singular matrices $\mathbf{V}$ and $\mathbf{W}$ such that $\mathbf{W}\mathbf{A}_2\mathbf{V}$, $\mathbf{W}\mathbf{A}_0\mathbf{V}$ are diagonal. Associated with system (1.106) is the generalized eigenvalue problem

$$\mathbf{A}_0\mathbf{v} = \psi\mathbf{A}_2\mathbf{v}. \tag{1.107}$$

This type of eigenvalue problem was considered by Ma and Caughey [26] while drawing upon the presentation of Zurmühl and Falk [78]. Associated with eigenvalue problem (1.107) is the adjoint eigenvalue problem

$$\mathbf{A}_0^T\mathbf{w} = \psi\mathbf{A}_2^T\mathbf{w}, \tag{1.108}$$

and the corresponding eigenvalues of (1.107) and (1.108) are identical because they lead to the same characteristic determinant.

Corresponding to each eigenvalue ψ_l, an eigenvector $\mathbf{v}_l$ can be found such that

$$\mathbf{A}_0\mathbf{v}_l = \psi_l\mathbf{A}_2\mathbf{v}_l. \tag{1.109}$$

A solution to the adjoint eigenvalue problem associated with the eigenvalue ψ_j is expressed by $\mathbf{A}_0^T\mathbf{w}_j = \psi_j\mathbf{A}_2^T\mathbf{w}_j$, and the transpose of this equation yields

$$\mathbf{w}_j^T\mathbf{A}_0 = \psi_j\mathbf{w}_j^T\mathbf{A}_2. \tag{1.110}$$

Pre-multiply Eq. (1.109) by $\mathbf{w}_j^T$ and post-multiply Eq. (1.110) by $\mathbf{v}_l$. Subtraction means that

$$(\psi_l - \psi_j)\mathbf{w}_j^T\mathbf{A}_2\mathbf{v}_l = 0. \tag{1.111}$$

The two sets of column vectors $\mathbf{v}_l$ and $\mathbf{w}_j$ are biorthogonal with respect to $\mathbf{A}_2$ if the corresponding eigenvalues are distinct. For simplicity, the eigenvalues are assumed distinct. However, Ma and Caughey [26] assumed that system (1.106) is non-defective. This presumes the independence of the eigenvectors while allowing repeated eigenvalues. The reader is referred to their paper for a treatment of repeated eigenvalues.

Assuming distinct eigenvalues, the biorthogonality relation can be expressed as

$$\mathbf{w}_l^T\mathbf{A}_2\mathbf{v}_j = \delta_{lj},\ l, j = 1,\ 2,\ \ldots,\ n. \tag{1.112}$$

In addition,

$$\mathbf{w}_l^T \mathbf{A}_0 \mathbf{v}_j = \psi_l \delta_{lj},\ l, j = 1,\ 2,\ \ldots,\ n. \tag{1.113}$$

Define the following square matrices of order n by

$$\mathbf{V} = \begin{bmatrix} \mathbf{v}_1 & \mathbf{v}_2 & \ldots & \mathbf{v}_n \end{bmatrix},\ \mathbf{W} = \begin{bmatrix} \mathbf{w}_1 & \mathbf{w}_2 & \ldots & \mathbf{w}_n \end{bmatrix}^T \tag{1.114}$$

so that the biorthogonality relations (1.112) and (1.113) can now be expressed as

$$\mathbf{W}\mathbf{A}_2\mathbf{V} = \mathbf{I},\ \mathbf{W}\mathbf{A}_0\mathbf{V} = \boldsymbol{\Omega}, \tag{1.115}$$

where $\boldsymbol{\Omega} = \text{diag}[\psi_1,\ \psi_2, \ldots,\ \psi_n]$. Now, let $\mathbf{q} = \mathbf{V}\mathbf{p}$ so Eq. (1.106) becomes

$$\ddot{\mathbf{p}} + \boldsymbol{\Omega}\mathbf{p} = \mathbf{W}\mathbf{f}(t). \tag{1.116}$$

Thus, the following theorem follows:

Theorem 1.8 *An undamped non-gyroscopic non-conservative system can always be decoupled by a common equivalence transformation.*

For a damped system (1.10), Eqs. (1.107) and (1.108) are identical and solution of only one of them is sufficient. The modes and adjoint modes are equal and $\mathbf{W} = \mathbf{V}^T = \mathbf{U}^T$, where $\mathbf{U}$ is the modal matrix. Thus, the decoupling equivalence transformation reduces to classical modal transformation if the coefficient matrices possess symmetry and positive-definiteness. The method of equivalence transformation represents a direct extension of classical modal analysis.

1.5.2 Non-conservative Systems

In classical modal analysis of damped systems, the modal matrix diagonalizes the mass matrix and stiffness matrix but will only diagonalize the damping matrix if and only if condition (1.56) is satisfied. Ma and Caughey [26] showed that a similar situation arises when decoupling non-conservative systems by equivalence transformations.

Referring to Eq. (1.11), an equivalence transformation diagonalizes $\mathbf{A}_2$ and $\mathbf{A}_0$. As before, let this equivalence transformation be defined by the two non-singular matrices $\mathbf{V}$ and $\mathbf{W}$. For $\mathbf{q} = \mathbf{V}\mathbf{p}$, Eq. (1.11) may be expressed in the form:

$$\ddot{\mathbf{p}} + \mathbf{W}\mathbf{A}_1\mathbf{V}\dot{\mathbf{p}} + \boldsymbol{\Omega}\mathbf{p} = \mathbf{W}\mathbf{f}(t). \tag{1.117}$$

It is observed that the condition

$$\mathbf{A}_1\mathbf{A}_2^{-1}\mathbf{A}_0 = \mathbf{A}_0\mathbf{A}_2^{-1}\mathbf{A}_1 \tag{1.118}$$

is equivalent to

$$\mathbf{D\Omega} = \mathbf{\Omega D}, \tag{1.119}$$

where $\mathbf{D} = \mathbf{WA}_1\mathbf{V}$. Condition (1.119) is satisfied if $\mathbf{D}$ is diagonal. Therefore, if the equivalence transformation decouples system (1.11), condition (1.118) is also satisfied.

On the contrary, assume that condition (1.118) is valid. Write $\mathbf{D} = [d_{lj}]$ and expand (1.119) to

$$\psi_j d_{lj} = \psi_l d_{lj}, \; l, j = 1, 2, \ldots, n. \tag{1.120}$$

With the assumption that all eigenvalues ψ_l are distinct, the above equation implies that $d_{lj} = 0$ when $l \neq j$. That means $\mathbf{D}$ is diagonal, and the equivalence transformation that diagonalizes $\mathbf{A}_2$ and $\mathbf{A}_0$ decouples the entire system. Consequently,

Theorem 1.9 *A non-conservative system can be decoupled by equivalence transformation if and only if* $\mathbf{A}_1\mathbf{A}_2^{-1}\mathbf{A}_0 = \mathbf{A}_0\mathbf{A}_2^{-1}\mathbf{A}_1$.

Condition (1.118) also holds for systems with semi-simple eigenvalues. In this case, the orthogonality conditions are obtained through an orthogonalization procedure. See [26] for detailed discussion. The process of decoupling by equivalence transformations is depicted in Fig. 1.10.

1.5.2.1 Example 1.12

The two-degree-of-freedom system (1.11) with coefficients

$$\mathbf{A}_2 = \begin{bmatrix} 2 & -1 \\ 1 & 0 \end{bmatrix}, \; \mathbf{A}_1 = \begin{bmatrix} 1 & 0 \\ 1 & 0 \end{bmatrix}, \; \mathbf{A}_0 = \begin{bmatrix} 1 & -1 \\ 0 & 0 \end{bmatrix} \tag{1.121}$$

is not diagonalizable by a similarity transformation, whereas it can be upper triangularized. In addition, it can be completely decoupled by equivalence transformation involving

$$\mathbf{V} = \begin{bmatrix} 1 & 0 \\ 1 & 1 \end{bmatrix}, \; \mathbf{W} = \begin{bmatrix} 0 & 1 \\ -1 & 1 \end{bmatrix}. \tag{1.122}$$

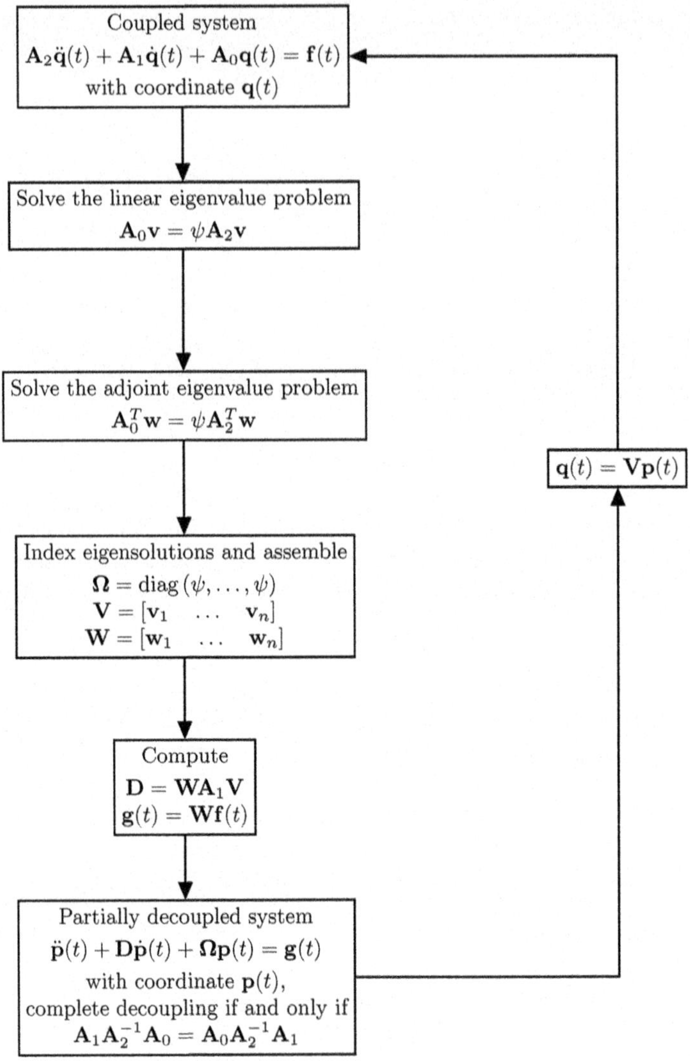

Fig. 1.10 Algorithm for decoupling a linear system by equivalence transformation. For classically damped systems, the modes and adjoint modes are equal so this procedure reduces to classical modal analysis and $\mathbf{W} = \mathbf{V}^T = \mathbf{U}^T$

For $\mathbf{q} = \mathbf{V}\mathbf{p}$,

$$\ddot{\mathbf{p}} + \begin{bmatrix} 1 & 0 \\ 0 & 0 \end{bmatrix} \dot{\mathbf{p}} + \begin{bmatrix} 0 & 0 \\ 0 & 1 \end{bmatrix} \mathbf{p} = \mathbf{W}\mathbf{f}(t). \tag{1.123}$$

1.5.2.2 Example 1.13

Consider the two-degree-of-freedom system:

$$\ddot{\mathbf{q}}+\begin{bmatrix}2 & 6\\ -8 & 8\end{bmatrix}\dot{\mathbf{q}}+\begin{bmatrix}0 & 3\\ -4 & 3\end{bmatrix}\mathbf{q}=\mathbf{f}(t). \tag{1.124}$$

The matrices:

$$\mathbf{V}=\frac{1}{8}\begin{bmatrix}3-i\sqrt{39} & 3+i\sqrt{39}\\ 8 & 8\end{bmatrix},$$

$$\mathbf{W}=\frac{1}{39}\begin{bmatrix}i4\sqrt{39} & \frac{39}{2}-i\frac{3\sqrt{39}}{2}\\ -i4\sqrt{39} & \frac{39}{2}+i\frac{3\sqrt{39}}{2}\end{bmatrix} \tag{1.125}$$

can be used to decouple this system by equivalence transformation using $\mathbf{q}=\mathbf{V}\mathbf{p}$:

$$\ddot{\mathbf{p}}+\begin{bmatrix}5+i\sqrt{39} & 0\\ 0 & 5-i\sqrt{39}\end{bmatrix}\dot{\mathbf{p}}$$
$$+\frac{1}{2}\begin{bmatrix}3+i\sqrt{39} & 0\\ 0 & 3-i\sqrt{39}\end{bmatrix}\mathbf{p}=\mathbf{W}\mathbf{f}(t). \tag{1.126}$$

In general, decoupling by equivalence transformations will lead to complex diagonal matrices, just as decoupling by similarity transformations. Similarly, such decoupling procedure lacks physical insight.

1.6 Conclusions

Decoupling the equation of motion of linear dynamical systems is a well-trodden problem that has attracted the attention of many researchers in the past century. Decoupling a damped system by modal analysis is equivalent to simultaneously diagonalizing three matrices by a congruence transformation. While modal analysis is the time-honored technique for decoupling classically damped systems, it cannot be used to advantage on non-conservative systems, whose coefficient matrices lack symmetry.

Equivalence transformations are the most general non-singular linear transformations and represent a direct extension of classical modal analysis. However, decoupling by equivalence transformation is also subject to mathematical restrictions. Mathematically speaking, no time-invariant linear transformations in the configuration space will decouple every linear system. Even partial decoupling, i.e., simultaneous conversion of $\mathbf{A}_2$, $\mathbf{A}_1$, and $\mathbf{A}_0$ into upper triangular forms, is not ensured with time-invariant linear transformations. As a consequence,

Remark 1.3 Any universal decoupling transformation in the configuration space must be at least time varying or even non-linear.

Part II
Method of Phase Synchronization

Chapter 2
Phase Synchronization and the Physics of Damping

To Thales . . . the primary question was not What do we know but How do we know it.

—Aristotle

A classically damped linear system possesses classical normal modes, which constitute a linear coordinate transformation that decouples the undamped system. This process of decoupling the equation of motion of a dynamical system is the time-honored procedure termed modal analysis. In general, damping is not classical and thus damped linear systems cannot be decoupled by modal analysis. This chapter shows how classical modal analysis can be extended to decouple damped linear systems in configuration space through a procedure called phase synchronization. Similar to modal analysis, decoupling by phase synchronization possesses ample physical insight based upon a consideration of the physics of damping. In this regard, phase angles are shifted in each non-classically damped mode of vibration so as to transform it into a classical mode. Eleven illustrative examples are provided.

2.1 Introduction

This chapter shows how modal analysis can be extended to decouple non-classically damped systems using the intuitive method of phase synchronization. It is argued how the method can be constructed by analyzing damped systems possessing distinct eigenvalues. This seems logical since classical engineering literature focuses mostly on passive systems.

This chapter is organized as follows. The difference between classically and non-classically damped modes is reviewed in Sect. 2.3. This difference elucidates how to construct time-shifting transformations for oscillatory modes in Sect. 2.3.2. Non-oscillatory damped motion is tackled in Sect. 2.4 and motion with mixed

R. Gonçalves Salsa Junior, F. Ma, *Advances in the Theory of System Decoupling*,
https://doi.org/10.1007/978-3-030-60846-0_2

modes in Sect. 2.5. Section 2.6 details the decoupling procedure for inhomogeneous equations. Section 2.7 shows how phase synchronization is related to modal analysis.

2.2 Damped Free Motion

The equation of motion of an n-degree-of-freedom damped linear system in free vibration can be written in the form

$$\mathbf{M}\ddot{\mathbf{q}} + \mathbf{C}\dot{\mathbf{q}} + \mathbf{K}\mathbf{q} = \mathbf{0}. \tag{2.1}$$

It is recalled that the coefficients **M**, **C**, and **K** are symmetric and positive-definite. The symmetric quadratic eigenvalue problem

$$(\lambda^2\mathbf{M} + \lambda\mathbf{C} + \mathbf{K})\mathbf{v} = \mathbf{0} \tag{2.2}$$

possesses $2n$ eigenvalues. Any complex eigenvalues and corresponding eigenvectors occur in complex conjugate pairs. Pre-multiply by the complex conjugate transpose $\mathbf{v}^*$ to obtain

$$\lambda^2(\mathbf{v}^*\mathbf{M}\mathbf{v}) + \lambda(\mathbf{v}^*\mathbf{C}\mathbf{v}) + (\mathbf{v}^*\mathbf{K}\mathbf{v}) = 0. \tag{2.3}$$

The roots of the above quadratic equation are given by

$$\lambda = \frac{-(\mathbf{v}^*\mathbf{C}\mathbf{v}) \pm \sqrt{(\mathbf{v}^*\mathbf{C}\mathbf{v})^2 - 4(\mathbf{v}^*\mathbf{M}\mathbf{v})(\mathbf{v}^*\mathbf{K}\mathbf{v})}}{2(\mathbf{v}^*\mathbf{M}\mathbf{v})}. \tag{2.4}$$

Due to positive-definiteness of the coefficients, the eigenvalues are either real and negative or complex with negative real part.

2.3 Oscillatory Damped Motion

For sake of simplicity, it will be assumed that the eigenvalues of the quadratic eigenvalue problem (2.2) are (i) complex and, in addition, (ii) distinct. These assumptions will be relaxed in the following sections. Moreover, it is guaranteed that all eigenvalues are complex if the discriminant

$$\Delta(\mathbf{v}) = (\mathbf{v}^*\mathbf{C}\mathbf{v})^2 - 4(\mathbf{v}^*\mathbf{M}\mathbf{v})(\mathbf{v}^*\mathbf{K}\mathbf{v}) < 0 \tag{2.5}$$

for arbitrary $\mathbf{v} \neq \mathbf{0}$ [59].

2.3.1 Damped Modes of Vibration

Assuming distinct eigenvalues, the real homogeneous solution may be expressed as [27]

$$\mathbf{q}(t) = \sum_{j=1}^{n}(c_j \mathbf{v}_j e^{\lambda_j t} + \bar{c}_j \bar{\mathbf{v}}_j e^{\bar{\lambda}_j t}) = \sum_{j=1}^{n} \mathbf{s}_j(t). \tag{2.6}$$

Thus $\mathbf{q}(t)$ is the superposition of n damped modes $\mathbf{s}_j(t)$. There are n damped modes of vibration and each damped mode is independently excitable; it should not be confused with an eigensolution $\mathbf{v}_j e^{\lambda_j t}$, which sometimes is referred to as complex mode [79]. A typical eigenvalue can be expressed as

$$\lambda_j = \alpha_j + i\omega_j \tag{2.7}$$

and its corresponding eigenvector can be written in polar form as

$$\mathbf{v}_j = \begin{bmatrix} r_{j1}e^{-i\phi_{j1}} \\ r_{j2}e^{-i\phi_{j2}} \\ \vdots \\ r_{jn}e^{-i\phi_{jn}} \end{bmatrix}. \tag{2.8}$$

Note that $\omega_j > 0, \alpha_j < 0$ and, if the phase drifts are taken as $\phi_{jl} \geq 0$, the amplitudes r_{jl} may be negative. For example, $e^{i\pi/4} = -e^{-i3\pi/4} = r_{jl}e^{-i\phi_{jl}}$. A damped mode of vibration is defined by pairing two complex conjugate eigensolutions in such a way that $\mathbf{s}_j(t) = c_j \mathbf{v}_j e^{\lambda_j t} + \bar{c}_j \bar{\mathbf{v}}_j e^{\bar{\lambda}_j t}$ becomes

$$\mathbf{s}_j(t) = \mathrm{Re}[2c_j \mathbf{v}_j e^{(\alpha_j + i\omega_j)t}] = A_j e^{\alpha_j t} \begin{bmatrix} r_{j1}\cos(\omega_j t - \phi_{j1} - \theta_j) \\ r_{j2}\cos(\omega_j t - \phi_{j2} - \theta_j) \\ \vdots \\ r_{jn}\cos(\omega_j t - \phi_{jn} - \theta_j) \end{bmatrix}, \tag{2.9}$$

where $2c_j$, an arbitrary multiplier, is written in polar forms as $2c_j = A_j e^{-i\theta_j}$. The $2n$ scalars A_j and θ_j $(j = 1, 2, \ldots, n)$ are determined from the initial conditions $\mathbf{q}(0)$ and $\dot{\mathbf{q}}(0)$ through

$$\mathbf{q}(0) = \sum_{j=1}^{n} A_j \begin{bmatrix} r_{j1}\cos(\phi_{j1} + \theta_j) \\ r_{j2}\cos(\phi_{j2} + \theta_j) \\ \vdots \\ r_{jn}\cos(\phi_{jn} + \theta_j) \end{bmatrix}, \tag{2.10}$$

and

$$\dot{\mathbf{q}}(0) = \sum_{j=1}^{n} \left\{ \alpha_j A_j \begin{bmatrix} r_{j1}\cos(\phi_{j1} + \theta_j) \\ r_{j2}\cos(\phi_{j2} + \theta_j) \\ \vdots \\ r_{jn}\cos(\phi_{jn} + \theta_j) \end{bmatrix} \right.$$

$$\left. + \omega_j A_j \begin{bmatrix} r_{j1}\sin(\phi_{j1} + \theta_j) \\ r_{j2}\sin(\phi_{j2} + \theta_j) \\ \vdots \\ r_{jn}\sin(\phi_{jn} + \theta_j) \end{bmatrix} \right\}. \tag{2.11}$$

A classically damped system satisfies condition (1.56). The eigenvalues of a classically damped system are associated with real eigenvectors $\mathbf{v}_j$, which coincide with mode shapes $\mathbf{u}_j$ of the symmetric undamped eigenvalue problem (1.40). As a consequence, both λ_j and $\bar{\lambda}_j$ share the same real eigenvector

$$\mathbf{v}_j = \bar{\mathbf{v}}_j = \mathbf{u}_j = \begin{bmatrix} r_{j1} \\ r_{j2} \\ \vdots \\ r_{jn} \end{bmatrix} \tag{2.12}$$

and, consequently, $\phi_{j1} = \phi_{j2} = \cdots = \phi_{jn} = 0$. In other words, two complex conjugate eigensolutions $\lambda_j = \alpha_j + i\omega_j$ and $\bar{\lambda}_j = \alpha_j - i\omega_j$ form a classically damped mode of vibration [17] with the functional form

$$\mathbf{s}_j(t) = \mathrm{Re}\left[2c_j\mathbf{u}_j e^{(\alpha_j + i\omega j)t}\right] = A_j e^{\alpha_j t}\cos(\omega_j t - \theta_j)\,\mathbf{u}_j. \tag{2.13}$$

Equation (2.13) represents synchronous motion in which all system components perform harmonic motion with the same damped frequency ω_j and exponential decay α_j, passing through their equilibrium positions at the same instant of time. Any two system components are either in phase or 180° out of phase with each other.

A particular case is that of an undamped system, for which the eigensolutions $\lambda_j = i\omega_j$ ($\alpha_j = 0$) combine to form a normal mode

$$\mathbf{s}_j(t) = \mathrm{Re}\left[2c_j\mathbf{u}_j e^{i\omega_j t}\right] = A_j\cos(\omega_j t - \theta_j)\,\mathbf{u}_j, \tag{2.14}$$

implying that the amplitude of modal vibration does not decrease and that ω_j is a natural frequency. A normal mode of vibration $\mathbf{s}_j(t) = A_j\cos(\omega_j t - \theta_j)\mathbf{u}_j$ represents synchronous motion in which all system components perform simple harmonic motion with the same frequency ω_j and the amplitude ratio of any two components is independent of time.

Classically damped systems possess a full set of classically damped modes, which are just the normal or natural modes with exponential decay. For convenience, no attempt will be made to distinguish between classically damped modes and normal modes.

In general, damped systems do not satisfy condition (1.56) and consequently they do not possess classically damped modes. In a non-classically damped system, the eigenvectors $\mathbf{v}_j$ are complex. A non-classically damped mode is physically excitable and has a well defined pattern, in which all system components perform harmonic vibration with identical frequency ω_j and with the same exponential decay α_j. However, elements of the modal vector $\mathbf{s}_j(t)$ in Eq. (2.9) generally possess different phase angles ϕ_{jl}. As a consequence, the system components do not pass through their equilibrium positions at the same time. On the other hand, the phase difference between any two elements in $\mathbf{s}_j(t)$ is constant. Thus the order in which the system components pass through their equilibrium positions remains unchanged. After one complete cycle of duration

$$T = \frac{2\pi}{\omega_j} \tag{2.15}$$

the components return to positions separated by the same phase angles as in the beginning of the cycle. There exists an unchanging pattern from cycle to cycle while the motion decays exponentially. Although the pattern is well defined, it is not as easy to recognize as a classical natural mode. Each non-classically damped mode $\mathbf{s}_j(t)$, for arbitrarily selected A_j and θ_j, can indeed be excited with the initial conditions [27]

$$\mathbf{q}(0) = A_j \begin{bmatrix} r_{j1}\cos(\phi_{j1}+\theta_j) \\ r_{j2}\cos(\phi_{j2}+\theta_j) \\ \vdots \\ r_{jn}\cos(\phi_{jn}+\theta_j) \end{bmatrix}, \tag{2.16}$$

and

$$\dot{\mathbf{q}}(0) = \alpha_j A_j \begin{bmatrix} r_{j1}\cos(\phi_{j1}+\theta_j) \\ r_{j2}\cos(\phi_{j2}+\theta_j) \\ \vdots \\ r_{jn}\cos(\phi_{jn}+\theta_j) \end{bmatrix} + \omega_j A_j \begin{bmatrix} r_{j1}\sin(\phi_{j1}+\theta_j) \\ r_{j2}\sin(\phi_{j2}+\theta_j) \\ \vdots \\ r_{jn}\sin(\phi_{jn}+\theta_j) \end{bmatrix}. \tag{2.17}$$

2.3.1.1 Example 2.1

Consider a system

$$\ddot{\mathbf{q}} + \mathbf{C}\dot{\mathbf{q}} + \begin{bmatrix} 2 & -1 \\ -1 & 2 \end{bmatrix} \mathbf{q} = \mathbf{0} \tag{2.18}$$

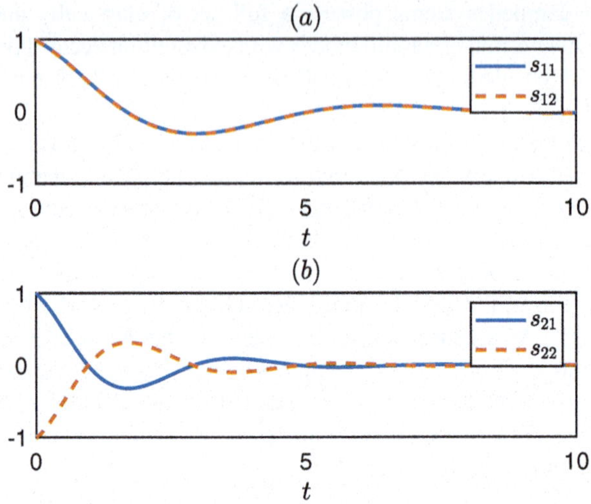

Fig. 2.1 System (2.18) is classically damped with damping matrix given by Eq. (2.19). (**a**) In the first classically damped mode both components oscillate with frequency $\omega_1 = \sqrt{55}/8$ and exponential decay $\alpha_1 = -3/8$ while both components are in phase. (**b**) In the second mode, $\omega_2 = \sqrt{167}/8$ and $\alpha_2 = -5/8$ and the system components are 180° out of phase with each other

for which

$$\mathbf{C} = \frac{1}{4}\begin{bmatrix} 4 & -1 \\ -1 & 4 \end{bmatrix}. \tag{2.19}$$

This system is classically damped and so it vibrates with classically damped modes of the form

$$\mathbf{s}_1(t) = \begin{bmatrix} s_{11}(t) \\ s_{12}(t) \end{bmatrix} = A_1 e^{-3t/8} \cos(\sqrt{55}t/8 - \theta_1) \begin{bmatrix} 1/\sqrt{2} \\ 1/\sqrt{2} \end{bmatrix}, \tag{2.20}$$

and

$$\mathbf{s}_2(t) = \begin{bmatrix} s_{21}(t) \\ s_{22}(t) \end{bmatrix} = A_2 e^{-5t/8} \cos(\sqrt{167}t/8 - \theta_2) \begin{bmatrix} 1/\sqrt{2} \\ -1/\sqrt{2} \end{bmatrix}. \tag{2.21}$$

These modes are illustrated in Fig. 2.1.

2.3.1.2 Example 2.2

Consider system (2.18) with $\mathbf{C} = \mathbf{0}$. Such system has normal modes given by

$$\mathbf{s}_1(t) = A_1 \cos(t - \theta_1) \begin{bmatrix} 1/\sqrt{2} \\ 1/\sqrt{2} \end{bmatrix}, \quad \mathbf{s}_2(t) = A_2 \cos(\sqrt{3}t - \theta_2) \begin{bmatrix} 1/\sqrt{2} \\ -1/\sqrt{2} \end{bmatrix}. \tag{2.22}$$

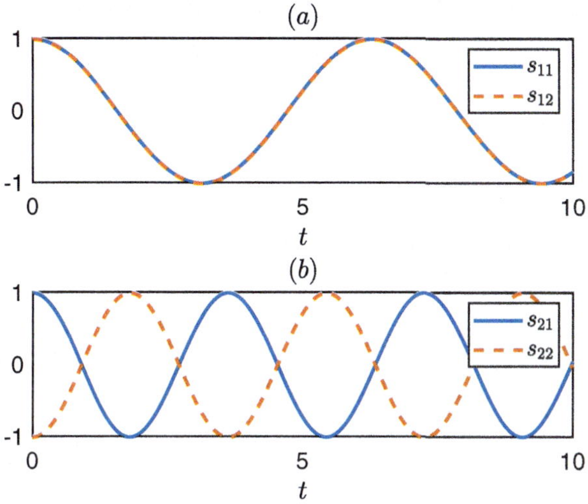

Fig. 2.2 System (2.18) with $\mathbf{C} = \mathbf{0}$ has normal modes (**a**) in phase with fundamental frequency $\omega_1 = 1$ and (**b**) 180° out of phase with frequency $\omega_2 = \sqrt{3}$

These modes are illustrated in Fig. 2.2. Indeed, classically damped modes are just exponentially decaying normal modes.

2.3.1.3 Example 2.3

System (2.18) with

$$\mathbf{C} = \frac{1}{10}\begin{bmatrix} 7 & -1 \\ -1 & 2 \end{bmatrix} \tag{2.23}$$

is non-classically damped because $\mathbf{KC} \neq \mathbf{CK}$; therefore it does not possess natural modes. However, damped modes can be computed. First note that the complex eigenvalues are

$$\lambda_1 = \bar{\lambda}_3 = \alpha_1 + i\omega_1 = -0.1792 + 1.0008i, \tag{2.24}$$

and

$$\lambda_2 = \bar{\lambda}_4 = \alpha_2 + i\omega_2 = -0.2708 + 1.6819i. \tag{2.25}$$

Normalized eigenvectors[1] can be written in polar form as

[1]Eigenvectors have been normalized in accordance with the scheme depicted in Appendix B.

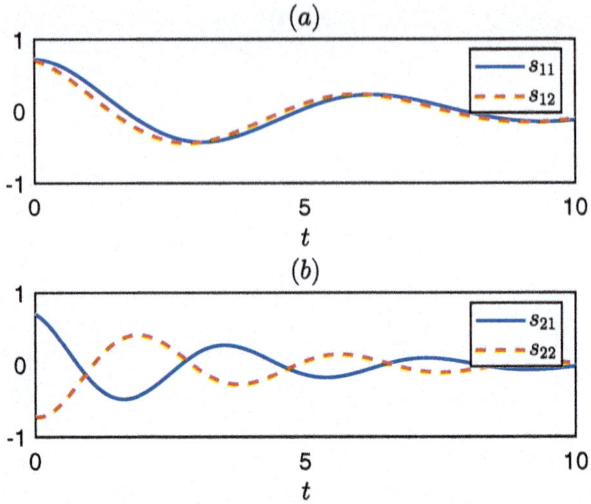

Fig. 2.3 System (2.18) is non-classically damped with damping matrix given by Eq. (2.23). (**a**) In the first damped mode, the systems oscillate with frequency $\omega_1 = 1.0008$ and exponential decay $\alpha_1 = -0.1792$. The component s_{11} lags behind s_{12} by 14.8740°. (**b**) In the second damped mode, $\omega_2 = 1.6819$ and $\alpha_2 = -0.2708$. The component s_{21} lags behind s_{22} by 154.4489°

$$\mathbf{v}_1 = \begin{bmatrix} 0.7389e^{-i7.3795°} \\ -0.7242e^{-i172.5055°} \end{bmatrix} = \begin{bmatrix} 0.7389e^{-i7.3795°} \\ 0.7242e^{i7.4945°} \end{bmatrix}, \tag{2.26}$$

and

$$\mathbf{v}_2 = \begin{bmatrix} -0.7336e^{-i167.1279°} \\ -0.7296e^{-i12.6790°} \end{bmatrix}. \tag{2.27}$$

The non-classically damped modes of vibration are

$$\mathbf{s}_1(t) = A_1 e^{-0.1792t} \begin{bmatrix} 0.7389\cos(1.0008t - 7.3795° - \theta_1) \\ 0.7242\cos(1.0008t + 7.4945° - \theta_1) \end{bmatrix}, \tag{2.28}$$

and

$$\mathbf{s}_2(t) = A_2 e^{-0.2708t} \begin{bmatrix} -0.7336\cos(1.6819t - 167.12792° - \theta_2) \\ -0.7296\cos(1.6819t - 12.6790° - \theta_2) \end{bmatrix}. \tag{2.29}$$

These modes are illustrated in Fig. 2.3. There is a constant phase difference of 14.8740° between the two elements in $\mathbf{s}_1(t)$. The constant phase difference is 154.4489° between the two elements of $\mathbf{s}_2(t)$.

2.3.2 Phase Synchronization of Complex Eigensolutions

While any two elements in a classically damped mode are either in phase or 180° out of phase with each other and they pass through their equilibrium positions at the same instant of time, the elements in a non-classically damped mode possess different phase angles, as given by ϕ_{jl}, and do not pass through their equilibrium positions at the same time. Thus the essential difference between a classically damped mode and a non-classically damped mode lies in the phase angles ϕ_{jl}. If suitable phase drifts are introduced in each non-classically damped mode so that all components are either in phase or out of phase, thus synchronizing the motion, it is possible to transform a non-classically damped system into one with classical damping and the resulting system can be decoupled using classical modal analysis. A basic objective of a process termed phase synchronization is to do just that. With the assumption of complex and distinct eigenvalues, Ma et al. [27] introduced phase shifts to the damped mode (2.9) so that

$$\mathbf{y}_j(t) = \begin{bmatrix} y_{j1}(t) \\ y_{j2}(t) \\ \vdots \\ y_{jn}(t) \end{bmatrix} = \begin{bmatrix} s_{j1}(t + \phi_{j1}/\omega_j) \\ s_{j2}(t + \phi_{j2}/\omega_j) \\ \vdots \\ s_{jn}(t + \phi_{jn}/\omega_j) \end{bmatrix}. \tag{2.30}$$

It follows that

$$\mathbf{y}_j(t) = A_j e^{\alpha_j t} \cos(\omega_j t - \theta_j) \begin{bmatrix} r_{j1} e^{\alpha_j \phi_{j1}/\omega_j} \\ r_{j2} e^{\alpha_j \phi_{j2}/\omega_j} \\ \vdots \\ r_{jn} e^{\alpha_j \phi_{jn}/\omega_j} \end{bmatrix}. \tag{2.31}$$

The vector $\mathbf{y}_j(t)$ can be simplified to

$$\mathbf{y}_j(t) = A_j e^{\alpha_j t} \cos(\omega_j t - \theta_j)\mathbf{z}_j = p_j(t)\mathbf{z}_j. \tag{2.32}$$

It has the same functional form as (2.13), so a system that possesses $\mathbf{y}_j(t)$ as its damped modes of vibrations is classically damped with $\mathbf{z}_j$ as its modal vectors. The effect of phase synchronization is to change the complex eigenvector $\mathbf{v}_j$ into a real mode shape $\mathbf{z}_j$. Each modal coordinate

$$p_j(t) = A_j e^{\alpha_j t} \cos(\omega_j t - \theta_j) = \mathrm{Re}[2c_j e^{\lambda_j t}] = c_j e^{\lambda_j t} + \bar{c}_j e^{\bar{\lambda}_j t} \tag{2.33}$$

of this classically damped system satisfies equations

$$\ddot{p}_j - (\lambda_j + \bar{\lambda}_j)\dot{p}_j + \lambda_j \bar{\lambda}_j p_j = 0, \tag{2.34}$$

which can be solved independently. The resulting decoupled system is easily created using only the eigenvalue information

$$\ddot{\mathbf{p}} + \mathbf{D}\dot{\mathbf{p}} + \boldsymbol{\Omega}\mathbf{p} = \mathbf{0}, \tag{2.35}$$

where the diagonal coefficients are given by

$$\mathbf{D} = -\begin{bmatrix} \lambda_1 + \bar{\lambda}_1 & & & \\ & \lambda_2 + \bar{\lambda}_2 & & \\ & & \ddots & \\ & & & \lambda_n + \bar{\lambda}_n \end{bmatrix}, \tag{2.36}$$

$$\boldsymbol{\Omega} = \begin{bmatrix} \lambda_1\bar{\lambda}_1 & & & \\ & \lambda_2\bar{\lambda}_2 & & \\ & & \ddots & \\ & & & \lambda_n\bar{\lambda}_n \end{bmatrix}. \tag{2.37}$$

What is the transformation that generates the original solution $\mathbf{q}(t)$ of the damped system (2.1) once the decoupled solution $\mathbf{p}(t)$ of (2.35) is obtained? Observe that the phase transformation (2.30) has an inverse such that

$$\mathbf{s}_j(t) = \begin{bmatrix} s_{j1}(t) \\ s_{j2}(t) \\ \vdots \\ s_{jn}(t) \end{bmatrix} = \begin{bmatrix} y_{j1}(t - \phi_{j1}/\omega_j) \\ y_{j2}(t - \phi_{j2}/\omega_j) \\ \vdots \\ y_{jn}(t - \phi_{jn}/\omega_j) \end{bmatrix}. \tag{2.38}$$

Using Eq. (2.32), Eq. (2.38) can be expressed as

$$\mathbf{s}_j(t) = \operatorname{diag}\left[p_j(t - \phi_{j1}/\omega_j), \ldots, p_j(t - \phi_{jn}/\omega_j)\right]\mathbf{z}_j. \tag{2.39}$$

As given in Eq. (2.6), $\mathbf{q}(t)$ is a superposition of the damped modes $\mathbf{s}_j(t)$. Thus,

$$\mathbf{q}(t) = \sum_{j=1}^{n} \operatorname{diag}\left[p_j(t - \phi_{j1}/\omega_j), \ldots, p_j(t - \phi_{jn}/\omega_j)\right]\mathbf{z}_j. \tag{2.40}$$

This coordinate transformation is real, linear, time-varying, and invertible. When $\mathbf{q}(t)$ is expanded as a superposition of modal vectors $\mathbf{u}_j$, decoupled equations for the modal coordinates cannot be obtained because a damped system generally does not vibrate in a normal mode. By phase synchronization, a classically damped system with modal vectors $\mathbf{z}_j$ is obtained. The response $\mathbf{q}(t)$ in free vibration can be expressed as a series in $\mathbf{z}_j$ with multipliers that are time-shifted modal coordinates of the tuned system.

By simply evaluating Eq. (2.40) and its derivative at $t = 0$, $\mathbf{q}$ and $\mathbf{p}$ are connected at different times. This happens because phase synchronization "pushes back" the element $s_{jl}(t)$ by an interval of length ϕ_{jl}/ω_j and consequently leads to a difficulty connecting the initial conditions at the same instant of time $t = 0$. This difficulty can be avoided if this transformation is cast as a state transformation. To do that, write

$$\mathbf{q}(t) = \sum_{j=1}^{n}(c_j\mathbf{v}_j e^{\lambda_j t} + \bar{c}_j\bar{\mathbf{v}}_j e^{\bar{\lambda}_j t}) = \mathbf{V}_1 e^{\mathbf{\Lambda}_1 t}\mathbf{c} + \mathbf{V}_2 e^{\mathbf{\Lambda}_2 t}\bar{\mathbf{c}} \tag{2.41}$$

where

$$\mathbf{V}_1 = \overline{\mathbf{V}}_2 = \begin{bmatrix}\mathbf{v}_1 & \mathbf{v}_2 & \cdots & \mathbf{v}_n\end{bmatrix}, \tag{2.42}$$

$$\mathbf{\Lambda}_1 = \bar{\mathbf{\Lambda}}_2 = \operatorname{diag}[\lambda_1, \lambda_2, \ldots, \lambda_n], \tag{2.43}$$

$$\mathbf{c} = \begin{bmatrix} c_1 \\ c_2 \\ \vdots \\ c_n \end{bmatrix}. \tag{2.44}$$

Since phase synchronization does not disturb the constants c_j in (2.41), it is possible to write the response $\mathbf{p}(t)$ of the decoupled system (2.35) as

$$\mathbf{p}(t) = e^{\mathbf{\Lambda}_1 t}\mathbf{c} + e^{\mathbf{\Lambda}_2 t}\bar{\mathbf{c}}. \tag{2.45}$$

As a state equation, these expressions become

$$\begin{bmatrix}\mathbf{q}(t) \\ \dot{\mathbf{q}}(t)\end{bmatrix} = \begin{bmatrix}\mathbf{V}_1 & \mathbf{V}_2 \\ \mathbf{V}_1\mathbf{\Lambda}_1 & \mathbf{V}_2\mathbf{\Lambda}_2\end{bmatrix}\begin{bmatrix}e^{\mathbf{\Lambda}_1 t} & \mathbf{0} \\ \mathbf{0} & e^{\mathbf{\Lambda}_2 t}\end{bmatrix}\begin{bmatrix}\mathbf{c} \\ \bar{\mathbf{c}}\end{bmatrix}, \tag{2.46}$$

and

$$\begin{bmatrix}\mathbf{p}(t) \\ \dot{\mathbf{p}}(t)\end{bmatrix} = \begin{bmatrix}\mathbf{I} & \mathbf{I} \\ \mathbf{\Lambda}_1 & \mathbf{\Lambda}_2\end{bmatrix}\begin{bmatrix}e^{\mathbf{\Lambda}_1 t} & \mathbf{0} \\ \mathbf{0} & e^{\mathbf{\Lambda}_2 t}\end{bmatrix}\begin{bmatrix}\mathbf{c} \\ \bar{\mathbf{c}}\end{bmatrix}. \tag{2.47}$$

Combine these state representations to yield

$$\begin{bmatrix}\mathbf{p}(t) \\ \dot{\mathbf{p}}(t)\end{bmatrix} = \begin{bmatrix}\mathbf{I} & \mathbf{I} \\ \mathbf{\Lambda}_1 & \mathbf{\Lambda}_2\end{bmatrix}\begin{bmatrix}\mathbf{V}_1 & \mathbf{V}_2 \\ \mathbf{V}_1\mathbf{\Lambda}_1 & \mathbf{V}_2\mathbf{\Lambda}_2\end{bmatrix}^{-1}\begin{bmatrix}\mathbf{q}(t) \\ \dot{\mathbf{q}}(t)\end{bmatrix} = \mathbf{S}\begin{bmatrix}\mathbf{q}(t) \\ \dot{\mathbf{q}}(t)\end{bmatrix}. \tag{2.48}$$

The state-space decoupling transformation is real, linear, time-invariant, invertible and, unlike a modal transformation, displacements and velocities can no longer be separated. One can readily obtain the initial values by evaluating Eq. (2.48) at $t = 0$.

Even though (2.48) is a state-space transformation, it is obtained through physical intuition with recourse to configuration-space analysis. The inverse of Eq. (2.48) is given by

$$\begin{aligned}\begin{bmatrix}\mathbf{q}(t)\\\dot{\mathbf{q}}(t)\end{bmatrix} &= \begin{bmatrix}\mathbf{V}_1 & \mathbf{V}_2\\\mathbf{V}_1\mathbf{\Lambda}_1 & \mathbf{V}_2\mathbf{\Lambda}_2\end{bmatrix}\begin{bmatrix}\mathbf{I} & \mathbf{I}\\\mathbf{\Lambda}_1 & \mathbf{\Lambda}_2\end{bmatrix}^{-1}\begin{bmatrix}\mathbf{p}(t)\\\dot{\mathbf{p}}(t)\end{bmatrix}\\&= \mathbf{T}\begin{bmatrix}\mathbf{p}(t)\\\dot{\mathbf{p}}(t)\end{bmatrix} = \begin{bmatrix}\mathbf{T}_1 & \mathbf{T}_2\\\mathbf{T}_3 & \mathbf{T}_4\end{bmatrix}\begin{bmatrix}\mathbf{p}(t)\\\dot{\mathbf{p}}(t)\end{bmatrix},\end{aligned} \tag{2.49}$$

where each real invertible submatrix $\mathbf{T}_j$ $(j = 1,\ 2,\ 3,\ 4)$ is of size $n \times n$. It can be shown that

$$\mathbf{T}_1 = (\mathbf{V}_1\mathbf{\Lambda}_2 - \mathbf{V}_2\mathbf{\Lambda}_1)\,(\mathbf{\Lambda}_2 - \mathbf{\Lambda}_1)^{-1}, \tag{2.50}$$

and

$$\mathbf{T}_2 = (\mathbf{V}_2 - \mathbf{V}_1)\,(\mathbf{\Lambda}_2 - \mathbf{\Lambda}_1)^{-1}. \tag{2.51}$$

This means

$$\mathbf{q}(t) = \left(\mathbf{T}_1 + \mathbf{T}_2\frac{d}{dt}\right)\mathbf{p}(t), \tag{2.52}$$

while modal transformation is simply $\mathbf{q} = \mathbf{U}\mathbf{p}$. Modal analysis amounts to finding a proper mix of displacements for decoupling in real space. Phase synchronization amounts to finding a proper mix of displacements and velocities for decoupling in real space.

2.3.2.1 Example 2.4

The damped system of Example 2.3 is non-classically damped and cannot be decoupled by modal analysis, as was illustrated in Example 1.4. Using phase synchronization, it can be decoupled into real single-degree-of-freedom systems by a real invertible transformation that is time-varying. Solution of the quadratic eigenvalue problem can be arranged as

$$\mathbf{\Lambda}_1 = \bar{\mathbf{\Lambda}}_2 = \begin{bmatrix}\lambda_1 & 0\\0 & \lambda_2\end{bmatrix} = \begin{bmatrix}-0.1792 + 1.0008i & 0\\0 & -0.2708 + 1.6819i\end{bmatrix} \tag{2.53}$$

and

$$\mathbf{V}_1 = \overline{\mathbf{V}}_2 = \begin{bmatrix}\mathbf{v}_1 & \mathbf{v}_2\end{bmatrix} = \begin{bmatrix}0.7328 - 0.0949i & 0.7152 + 0.1634i\\0.7180 + 0.0945i & -0.7118 + 0.1601i\end{bmatrix}. \tag{2.54}$$

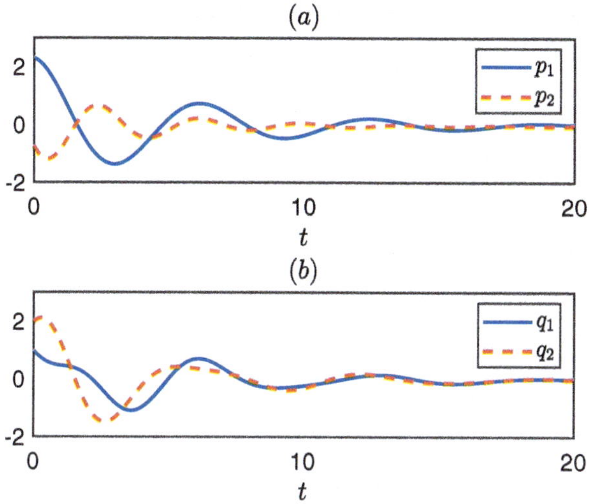

Fig. 2.4 (**a**) Response of decoupled system and (**b**) response of coupled system of Example 2.4. The initial conditions used are $\mathbf{q}(0) = [1 \quad 2]^T$ and $\dot{\mathbf{q}}(0) = [-1 \quad 1]^T$, which result in $\mathbf{p}(0) = [2.3221 \quad -0.7111]^T$ and $\dot{\mathbf{p}}(0) = [-0.3318 \quad -1.7120]^T$. While the homogeneous part of the decoupled system is constructed from the eigenvalues, the decoupling transformation and initial conditions of the decoupled system depend on the eigenvectors and thus on the normalization scheme used

The eigenvectors have been normalized. The decoupled system is given by

$$\ddot{\mathbf{p}} + \mathbf{D}\dot{\mathbf{p}} + \boldsymbol{\Omega}\mathbf{p} = \ddot{\mathbf{p}} - (\boldsymbol{\Lambda}_1 + \boldsymbol{\Lambda}_2)\dot{\mathbf{p}} + \boldsymbol{\Lambda}_1\boldsymbol{\Lambda}_2\mathbf{p} = \mathbf{0}, \tag{2.55}$$

which results in

$$\ddot{\mathbf{p}} + \begin{bmatrix} 0.3584 & 0 \\ 0 & 0.5416 \end{bmatrix} \dot{\mathbf{p}} + \begin{bmatrix} 1.0337 & 0 \\ 0 & 2.9022 \end{bmatrix} \mathbf{p} = \mathbf{0}. \tag{2.56}$$

The responses of the coupled and decoupled systems are illustrated in Fig. 2.4.

Analytical formulas for the decoupling transformation can be readily constructed. Compute the real matrix

$$\mathbf{S} = \begin{bmatrix} 0.6740 & 0.7294 & -0.0948 & 0.0944 \\ 0.7474 & -0.7282 & 0.0972 & 0.0952 \\ 0.2840 & -0.2836 & 0.7498 & 0.7011 \\ -0.0991 & -0.0932 & 0.6889 & -0.7376 \end{bmatrix}. \tag{2.57}$$

The state-space decoupling transformation is given by $\mathbf{p} = \mathbf{S}\mathbf{q}$, which results in

$$\begin{bmatrix} p_1 \\ p_2 \\ \dot{p}_1 \\ \dot{p}_2 \end{bmatrix} = \begin{bmatrix} 0.6740 & 0.7294 & -0.0948 & 0.0944 \\ 0.7474 & -0.7282 & 0.0972 & 0.0952 \\ 0.2840 & -0.2836 & 0.7498 & 0.7011 \\ -0.0991 & -0.0932 & 0.6889 & -0.7376 \end{bmatrix} \begin{bmatrix} q_1 \\ q_2 \\ \dot{q}_1 \\ \dot{q}_2 \end{bmatrix}. \tag{2.58}$$

A configuration-space transformation can be obtained by

$$
\begin{aligned}
\mathbf{T} = \mathbf{S}^{-1} &= \begin{bmatrix} \mathbf{T}_1 & \mathbf{T}_2 \\ \mathbf{T}_3 & \mathbf{T}_4 \end{bmatrix} \\
&= \begin{bmatrix} 0.7158 & 0.7415 & -0.0948 & 0.0972 \\ 0.7349 & -0.6860 & 0.0944 & 0.0952 \\ 0.0980 & -0.2820 & 0.7498 & 0.6889 \\ -0.0976 & -0.2763 & 0.7011 & -0.7376 \end{bmatrix},
\end{aligned} \tag{2.59}
$$

where each submatrix $\mathbf{T}_j$ $(j = 1,\ 2,\ 3,\ 4)$ is of dimension 2×2. This means

$$
\begin{aligned}
\mathbf{q} &= \left(\mathbf{T}_1 + \mathbf{T}_2 \frac{d}{dt}\right)\mathbf{p} \\
&= \begin{bmatrix} 0.7158 & 0.7415 \\ 0.7349 & -0.6860 \end{bmatrix} \begin{bmatrix} p_1 \\ p_2 \end{bmatrix} + \begin{bmatrix} -0.0948 & 0.0972 \\ 0.0944 & 0.0952 \end{bmatrix} \begin{bmatrix} \dot{p}_1 \\ \dot{p}_2 \end{bmatrix}.
\end{aligned} \tag{2.60}
$$

Alternatively, a decoupling transformation as a linear time-shifting transformation can be constructed. The modal matrix of the classically damped system that arises from phase synchronization is

$$
\begin{aligned}
\mathbf{Z} = \begin{bmatrix} \mathbf{z}_1 & \mathbf{z}_2 \end{bmatrix} &= \begin{bmatrix} r_{11} e^{\alpha_1 \phi_{11}/\omega_1} & r_{21} e^{\alpha_2 \phi_{21}/\omega_2} \\ r_{12} e^{\alpha_1 \phi_{12}/\omega_1} & r_{22} e^{\alpha_2 \phi_{22}/\omega_2} \end{bmatrix} \\
&= \begin{bmatrix} 0.7221 & -0.4587 \\ -0.4224 & -0.7040 \end{bmatrix}.
\end{aligned} \tag{2.61}
$$

After phase synchronization, the damped modes of vibration are given by Eq. (2.32):

$$
\mathbf{y}_1(t) = A_1 e^{-0.1792t} \cos(1.0008t - \theta_1) \begin{bmatrix} 0.7221 \\ -0.4224 \end{bmatrix} \tag{2.62}
$$

and

$$
\mathbf{y}_2(t) = A_2 e^{-0.2708t} \cos(1.6819t - \theta_2) \begin{bmatrix} -0.4587 \\ -0.7040 \end{bmatrix}. \tag{2.63}
$$

The decoupling transformation, as a linear time-shifting transformation, is given by Eq. (2.40)

$$\mathbf{q}(t) = \begin{bmatrix} p_1(t - \phi_{11}/\omega_1) & 0 \\ 0 & p_1(t - \phi_{12}/\omega_1) \end{bmatrix} \mathbf{z}_1 + \begin{bmatrix} p_2(t - \phi_{21}/\omega_2) & 0 \\ 0 & p_2(t - \phi_{22}/\omega_2) \end{bmatrix} \mathbf{z}_2, \tag{2.64}$$

which results in

$$\mathbf{q}(t) = \begin{bmatrix} p_1(t - 0.1287) & 0 \\ 0 & p_1(t - 3.0084) \end{bmatrix} \begin{bmatrix} 0.7221 \\ -0.4224 \end{bmatrix} + \begin{bmatrix} p_2(t - 1.7343) & 0 \\ 0 & p_2(t - 0.1316) \end{bmatrix} \begin{bmatrix} -0.4587 \\ -0.7040 \end{bmatrix}, \tag{2.65}$$

In comparison, there are no time drifts in modal transformations.

2.3.3 *Uniqueness of Decoupled System*

Does phase synchronization generate a unique decoupled system represented by Eq. (2.35)? Two decoupled systems are regarded as the same if their component equations coincide; the order in which the component equations appear is immaterial. Hence, indexing schemes for the eigensolutions that re-order the component equations of Eq. (2.35) generate equivalent systems. For a given indexing scheme, the coefficient matrices $\mathbf{D}$ and $\boldsymbol{\Omega}$ of Eq. (2.35) are independent of the normalization of eigenvectors because they are constructed from the eigenvalues. On the other hand, the initial values of the decoupled system and decoupling transformation depend on the eigenvectors and hence on their normalization schemes.

It is usual to index the eigenvalues with positive imaginary parts in order of increasing magnitude of their imaginary parts as the first n eigenvalues such that

$$0 < \mathrm{Im}[\lambda_1] \le \mathrm{Im}[\lambda_2] \le \ldots \le \mathrm{Im}[\lambda_n]. \tag{2.66}$$

Since $\mathrm{Im}[\lambda_j]$ can be regarded as a frequency of vibration, this is consistent with the convention of arranging frequencies in order of increasing magnitude. The remaining eigenvalues, which are the complex conjugates with negative imaginary parts, can be conveniently enumerated as $\lambda_{n+1} = \bar{\lambda}_1, \lambda_{n+2} = \bar{\lambda}_2, \ldots, \lambda_{2n} = \bar{\lambda}_n$.

2.3.3.1 Example 2.5

To illustrate the effect of normalization of eigenvectors, consider system of Example 2.4 but suppose now that the eigenvectors are normalized in order that each state eigenvector $[\mathbf{v}_j \quad \lambda_j \mathbf{v}_j]^T$ has unit Euclidean norm. In this case, the decoupled system (2.56) remains the same, but the initial values now are

$$\begin{bmatrix} p_1(0) \\ p_2(0) \end{bmatrix} = \begin{bmatrix} -0.0416 \\ -1.6007 \end{bmatrix}, \quad \begin{bmatrix} \dot{p}_1(0) \\ \dot{p}_2(0) \end{bmatrix} = \begin{bmatrix} -3.4234 \\ 4.1588 \end{bmatrix}. \tag{2.67}$$

The decoupling transformation becomes

$$\mathbf{q} = \begin{bmatrix} -0.1765 & -0.1141 \\ -0.0472 & 0.2505 \end{bmatrix} \begin{bmatrix} p_1 \\ p_2 \end{bmatrix} + \begin{bmatrix} -0.4926 & -0.2107 \\ -0.4888 & 0.1745 \end{bmatrix} \begin{bmatrix} \dot{p}_1 \\ \dot{p}_2 \end{bmatrix}. \tag{2.68}$$

2.4 Non-oscillatory Damped Motion

The assumption that the eigenvalues are complex is now relaxed. Consider purely non-oscillatory free motion of (2.1) for which the $2n$ eigenvalues of Eq. (2.2) are (i) real and (ii) distinct. The corresponding eigenvectors can be chosen to be real and, as a consequence, all eigensolutions are taken as real.

In Sect. 2.3, two complex conjugate eigensolutions are paired to generate an underdamped mode of vibration. In contrast, any two distinct real eigensolutions may be combined to generate an overdamped mode of vibration. The damped mode $\mathbf{s}_j(t)$ can be defined by pairing the eigensolution $\mathbf{v}_j e^{\lambda_j t}$ with a distinct eigensolution, say $\mathbf{v}_k e^{\lambda_k t}$, in a linear combination such that

$$\mathbf{s}_j(t) = c_j \mathbf{v}_j e^{\lambda_j t} + c_k \mathbf{v}_k e^{\lambda_k t}, \tag{2.69}$$

where c_j and c_k are two arbitrary real constants. A typical element of $\mathbf{s}_j(t)$ has the form

$$s_{jl}(t) = c_j v_{jl} e^{\lambda_j t} + c_k v_{kl} e^{\lambda_k t}. \tag{2.70}$$

Thus each element of $\mathbf{s}_j(t)$ is non-oscillatory and exponentially decaying, analogous to an overdamped free response.

In Sect. 2.3.2, a non-classically damped system is transformed into a classically damped system by tuning the damped modes. Can phase synchronization, applicable to oscillatory free vibration, be modified to decouple non-oscillatory free vibration? What are the phase angles to synchronize?

2.4.1 *Vibration with Imaginary Frequency*

Consider a single-degree-of-freedom overdamped system of the form

$$m\ddot{q} + c\dot{q} + kq = 0. \tag{2.71}$$

This is a realization of Eq. (2.1) with $n = 1$, for which the coefficients are positive numbers. The two real and distinct eigenvalues are

$$\lambda_1,\ \lambda_2 = \frac{1}{2m}(-c \mp \sqrt{c^2 - 4mk}), \tag{2.72}$$

where $c^2 - 4mk > 0$. The general solution is

$$q(t) = c_1 e^{\lambda_1 t} + c_2 e^{\lambda_2 t} \tag{2.73}$$

and c_1 and c_2 are determined from initial conditions.

From a strict mathematical viewpoint, there need not be day and night distinction between oscillatory and non-oscillatory behaviors. Real exponential decay of a damped system can be regarded as oscillations with imaginary frequency. To this end, the concept of real quadratic conjugation is introduced [28].

Were these two eigenvalues complex, they would be complex conjugates of each other. Since they are real, they may be referred to as real quadratic conjugates using the notation $\lambda_1 = \tilde{\lambda}_2$. A pair of real quadratic conjugates are the two real roots of a common quadratic equation. In order to take advantage of the algebra of complex variables, real quadratic conjugates will be expressed in complex notation, such as rectangular and polar forms. Define

$$\alpha = \frac{1}{2}(\lambda_1 + \lambda_2) = -\frac{c}{2m} \tag{2.74}$$

and

$$\omega = \frac{i}{2}|\lambda_1 - \lambda_2| = \frac{i}{2m}\sqrt{c^2 - 4mk}. \tag{2.75}$$

This means $\lambda_1 = \alpha + i\omega$ and $\lambda_2 = \tilde{\lambda}_1 = \alpha - i\omega$. Note the change in notation

$$\widetilde{\mathrm{Re}}[\lambda_1] = \alpha, \quad \widetilde{\mathrm{Im}}[\lambda_1] = \omega \tag{2.76}$$

because the usual notation results in

$$\mathrm{Re}[\lambda_1] = \lambda_1, \quad \mathrm{Im}[\lambda_1] = 0. \tag{2.77}$$

The real coefficients c_1 and c_2 are two real numbers which may be regarded as a pair of real quadratic conjugates so that $c_1 = \tilde{c}_2$ and the response (2.73) can be recast as

$$q(t) = \widetilde{\mathrm{Re}}[2c_1 e^{\alpha + i\omega}]. \tag{2.78}$$

Let $2c_1 = Ae^{-i\theta}$ so that $q(t)$ is expressed as

$$q(t) = Ae^{\alpha}\cos(\omega t - \theta), \tag{2.79}$$

where

$$A,\ \theta = \begin{cases} 2\sqrt{c_1 c_2},\ \frac{i}{2}\ln\frac{c_1}{c_2} & \text{if } c_1 > 0,\ c_2 > 0. \\ -2\sqrt{|c_1 c_2|},\ \frac{i}{2}\ln\left|\frac{c_1}{c_2}\right| & \text{if } c_1 < 0,\ c_2 < 0. \\ 2i\sqrt{|c_1 c_2|},\ -\frac{\pi}{2} + \frac{i}{2}\ln\left|\frac{c_1}{c_2}\right| & \text{if } c_1 < 0,\ c_2 > 0. \\ 2i\sqrt{|c_1 c_2|},\ \frac{\pi}{2} + \frac{i}{2}\ln\left|\frac{c_1}{c_2}\right| & \text{if } c_1 > 0,\ c_2 < 0. \end{cases} \tag{2.80}$$

The values in Eq. (2.80) can be readily checked. When $c_1 > 0$ and $c_2 > 0$,

$$\begin{aligned} q(t) &= A\, e^{\alpha t}\, (e^{i(\omega t - \theta)} + e^{-i(\omega t - \theta)})/2 \\ &= \sqrt{c_1 c_2}\left[e^{(\alpha + i\omega)t} e^{\frac{1}{2}\ln\frac{c_1}{c_2}} + e^{(\alpha - i\omega)t} e^{-\frac{1}{2}\ln\frac{c_1}{c_2}}\right] \\ &= \sqrt{c_1 c_2}\left[e^{\lambda_1 t}\sqrt{\frac{c_1}{c_2}} + e^{\lambda_2 t}\sqrt{\frac{c_2}{c_1}}\right] = c_1 e^{\lambda_1 t} + c_2 e^{\lambda_2 t}. \end{aligned} \tag{2.81}$$

In a similar fashion, it can be shown that when $c_1 < 0$ and $c_2 < 0$,

$$q(t) = -|c_1| e^{\lambda_1 t} - |c_2| e^{\lambda_2 t}. \tag{2.82}$$

When $c_1 < 0$ and $c_2 > 0$,

$$\begin{aligned} q(t) &= i\sqrt{|c_1 c_2|}\left[e^{\lambda_1 t}\sqrt{\left|\frac{c_1}{c_2}\right|}\, e^{i\pi/2} + e^{\lambda_2 t}\sqrt{\left|\frac{c_2}{c_1}\right|}\, e^{-i\pi/2}\right] \\ &= -|c_1| e^{\lambda_1 t} + |c_2| e^{\lambda_2 t}. \end{aligned} \tag{2.83}$$

Similarly, when $c_1 > 0$ and $c_2 < 0$,

$$q(t) = |c_1| e^{\lambda_1 t} - |c_2| e^{\lambda_2 t}. \tag{2.84}$$

Thus real exponential decay has been represented as oscillations with imaginary frequency, for which amplitude and phase angle may not be real. These results may be described in terms of hyperbolic functions.

Since imaginary vibration exhibits itself physically as real exponential decay, its characterization can be subtle at times and real quadratic conjugates should be interpreted appropriately in limiting cases. For example, when either $c_1 = 0$ or $c_2 = 0$, the amplitude A and phase angle θ result in an indeterminate form. This can be evaluated by recognizing that $q(t)$ represents real exponential decay with $q(t) = q(0)e^{(\alpha+i\omega)t}$ or $q(t) = q(0)e^{(\alpha-i\omega)t}$ [28, 30].

2.4.1.1 Example 2.6

Consider the overdamped system

$$\ddot{q} + 4\dot{q} + 3q = 0 \tag{2.85}$$

with $q(0) = 3$ and $\dot{q}(0) = -5$. The eigenvalues are $\lambda_1 = -3$ and $\lambda_2 = -1$; the system response is

$$q(t) = e^{-3t} + 2e^{-t}. \tag{2.86}$$

Such system can be thought of as vibrating with imaginary frequency. This is achieved through the following steps:

1. The eigenvalues may be rewritten as $\lambda_1 = \alpha + i\omega$ and $\lambda_2 = \alpha - i\omega = \tilde{\lambda}_1$, where

$$\alpha = -2, \ \omega = i. \tag{2.87}$$

2. Since $c_1 = 1$ and $c_2 = 2$,

$$A = \sqrt{c_1 c_2} = \sqrt{2} \tag{2.88}$$

and

$$\theta = \frac{i}{2}\ln\frac{c_1}{c_2} = \frac{i}{2}\ln\frac{1}{2}. \tag{2.89}$$

The solution is then

$$q(t) = Ae^{\alpha}\cos(\omega t - \theta) = 2\sqrt{2}\, e^{-2t} \cosh\left(t + \frac{\ln(2)}{2}\right). \tag{2.90}$$

Equation (2.90) simplifies to Eq. (2.86) using the relation

$$\cosh\left(t + \frac{\ln(2)}{2}\right) = \frac{\sqrt{2}}{4}e^{-t} + \frac{\sqrt{2}}{2}e^{t}. \tag{2.91}$$

2.4.2 *Phase Synchronization of Real Eigensolutions*

Phase synchronization can still be applied to non-oscillatory damped modes (2.69) through the concept of vibration with imaginary frequency [28, 30]. The real eigenvalues λ_j and λ_k can be regarded as roots of a quadratic factor of the characteristic polynomial associated with (2.2). Hence,

$$\lambda_j,\ \lambda_k = \alpha_j \pm i\omega_j \tag{2.92}$$

with

$$\alpha_j = \frac{1}{2}(\lambda_j + \lambda_k) \tag{2.93}$$

and

$$\omega = \frac{i}{2}|\lambda_j - \lambda_k|. \tag{2.94}$$

The two eigensolutions are considered as real quadratic conjugates and, inspired by Eq. (2.80), the corresponding eigenvectors can be expressed as

$$\mathbf{v}_j = \begin{bmatrix} v_{j1} \\ v_{j2} \\ \vdots \\ v_{jn} \end{bmatrix} = \begin{bmatrix} r_{j1}e^{-i\phi_{j1}} \\ r_{j2}e^{-i\phi_{j2}} \\ \vdots \\ r_{jn}e^{-i\phi_{jn}} \end{bmatrix},\ \mathbf{v}_k = \tilde{\mathbf{v}}_j = \begin{bmatrix} v_{k1} \\ v_{k2} \\ \vdots \\ v_{kn} \end{bmatrix} = \begin{bmatrix} r_{j1}e^{i\phi_{j1}} \\ r_{j2}e^{i\phi_{j2}} \\ \vdots \\ r_{jn}e^{i\phi_{jn}} \end{bmatrix}, \tag{2.95}$$

where

$$r_{jl},\ \phi_{jl} = \begin{cases} \sqrt{v_{jl}v_{kl}},\ \dfrac{i}{2}\ln\dfrac{v_{jl}}{v_{kl}} & \text{if } v_{jl} > 0,\ v_{kl} > 0. \\ -\sqrt{|v_{jl}v_{kl}|},\ \dfrac{i}{2}\ln\left|\dfrac{v_{jl}}{v_{kl}}\right| & \text{if } v_{jl} < 0,\ v_{kl} < 0. \\ i\sqrt{|v_{jl}v_{kl}|},\ -\dfrac{\pi}{2} + \dfrac{i}{2}\ln\left|\dfrac{v_{jl}}{v_{kl}}\right| & \text{if } v_{jl} < 0,\ v_{kl} > 0. \\ i\sqrt{|v_{jl}v_{kl}|},\ \dfrac{\pi}{2} + \dfrac{i}{2}\ln\left|\dfrac{v_{jl}}{v_{kl}}\right| & \text{if } v_{jl} > 0,\ v_{kl} < 0. \end{cases} \tag{2.96}$$

If $2c_1 = 2\tilde{c}_2 = A_j e^{-\theta_j}$, for which A_j and θ_j are given by Eq. (2.80), the mode (2.69) can then be recast as

$$\mathbf{s}_j(t) = \widetilde{\mathrm{Re}}[2c_j\mathbf{v}_j e^{(\alpha_j + i\omega_j)t}] = A_j e^{\alpha_j t} \begin{bmatrix} r_{j1}\cos(\omega_j t - \phi_{j1} - \theta_j) \\ r_{j2}\cos(\omega_j t - \phi_{j2} - \theta_j) \\ \vdots \\ r_{jn}\cos(\omega_j t - \phi_{jn} - \theta_j) \end{bmatrix}, \tag{2.97}$$

which has the same functional form as an underdamped mode generated by two complex conjugate eigensolutions. The two real numbers v_{jl} and v_{kl} constitute a pair of real quadratic conjugates; the amplitude r_{jl} is either real or imaginary and the phase angle ϕ_{jl} is either imaginary or complex with a constant real part, even though $\mathbf{s}_j(t)$ is always real and can be physically excited with the real initial conditions

$$\mathbf{q}(0) = c_j \mathbf{v}_j + \tilde{c}_j \tilde{\mathbf{v}}_j = A_j \begin{bmatrix} r_{j1}\cos(\phi_{j1} + \theta_j) \\ r_{j2}\cos(\phi_{j2} + \theta_j) \\ \vdots \\ r_{jn}\cos(\phi_{jn} + \theta_j) \end{bmatrix} \tag{2.98}$$

and

$$\begin{aligned} \dot{\mathbf{q}}(0) &= c_j \lambda_j \mathbf{v}_j + \tilde{c}_j \tilde{\lambda}_j \tilde{\mathbf{v}}_j \\ &= \alpha_j A_j \begin{bmatrix} r_{j1}\cos(\phi_{j1} + \theta_j) \\ r_{j2}\cos(\phi_{j2} + \theta_j) \\ \vdots \\ r_{jn}\cos(\phi_{jn} + \theta_j) \end{bmatrix} + \omega_j A_j \begin{bmatrix} r_{j1}\sin(\phi_{j1} + \theta_j) \\ r_{j2}\sin(\phi_{j2} + \theta_j) \\ \vdots \\ r_{jn}\sin(\phi_{jn} + \theta_j) \end{bmatrix}. \end{aligned} \tag{2.99}$$

In the case v_{jl} or v_{kl} equals zero, no extra work is required and $s_{jl}(t)$ is either $c_j v_{jl} e^{\lambda_j t}$ or $c_k v_{kl} e^{\lambda_k t}$. The same applies if either c_j or c_k equals zero [28, 30].

Oscillatory free vibration has no divine rights in its functional representation as a superposition of sinusoidal oscillations. Every formula established for oscillatory free vibration can be suitably applied to purely non-oscillatory free vibration with the use of imaginary frequencies. Subsequently, phase synchronization operates in the same way as for oscillatory damped modes, amounting to introducing complex phase shifts to each mode $\mathbf{s}_j(t)$ so that a classically damped mode is generated

$$\begin{aligned} \mathbf{y}_j(t) &= \begin{bmatrix} s_{j1}(t + \phi_{j1}/\omega_j) \\ s_{j2}(t + \phi_{j2}/\omega_j) \\ \vdots \\ s_{jn}(t + \phi_{jn}/\omega_j) \end{bmatrix} = A_j e^{\alpha_j t} \cos(\omega_j t - \theta_j) \begin{bmatrix} r_{j1} e^{\alpha_j \phi_{j1}/\omega_j} \\ r_{j2} e^{\alpha_j \phi_{j2}/\omega_j} \\ \vdots \\ r_{jn} e^{\alpha_j \phi_{jn}/\omega_j} \end{bmatrix} \\ &= A_j e^{\alpha_j t} \cos(\omega_j t - \theta_j) \mathbf{z}_j = p_j(t) \mathbf{z}_j \end{aligned} \tag{2.100}$$

and the corresponding decoupled system has diagonal coefficients

$$\mathbf{D} = - \begin{bmatrix} \lambda_1 + \tilde{\lambda}_1 & & & \\ & \lambda_2 + \tilde{\lambda}_2 & & \\ & & \ddots & \\ & & & \lambda_n + \tilde{\lambda}_n \end{bmatrix}, \tag{2.101}$$

$$\boldsymbol{\Omega} = \begin{bmatrix} \lambda_1\tilde{\lambda}_1 & & & \\ & \lambda_2\tilde{\lambda}_2 & & \\ & & \ddots & \\ & & & \lambda_n\tilde{\lambda}_n \end{bmatrix}. \tag{2.102}$$

When v_{jl} and v_{kl} have the same sign, ϕ_{jl} is imaginary and r_{jl} is real. Thus if corresponding elements of $\mathbf{v}_j$ and $\mathbf{v}_k$ have the same sign, all time shifts ϕ_{jl}/ω_j and tuned mode shapes $\mathbf{z}_j$ are real. Any imaginary part that arises in the time shift is just to accommodate a change in sign between v_{jl} and v_{kl}.

A system possessing the synchronized modes $\mathbf{y}_j(t)$ is classically damped. Basically, all equations for decoupling oscillatory free vibration can be carried over to non-oscillatory free vibration provided that complex conjugation is suitably replaced by real quadratic conjugation. Specifically,

$$\mathbf{q}(t) = \begin{bmatrix} \sum_{j=1}^{n} p_j(t - \phi_{j1}/\omega_j) z_{j1} \\ \sum_{j=1}^{n} p_j(t - \phi_{j2}/\omega_j) z_{j2} \\ \vdots \\ \sum_{j=1}^{n} p_j(t - \phi_{jn}/\omega_j) z_{jn} \end{bmatrix}. \tag{2.103}$$

The right-hand side of the above equation is indeed real and equal to $\sum_{j=1}^{n} \mathbf{s}_j(t)$. To see that, note $p_j(t) = c_j e^{\lambda_j t} + c_k e^{\lambda_k t}$ and that a typical element on the right-hand side of the above expression is

$$\begin{aligned} p_j(t - \phi_{jl}/\omega_j) z_{jl} &= r_{jl}[c_j e^{\lambda_j t} e^{-i\phi_{jl}} + c_k e^{\lambda_k t} e^{i\phi_{jl}}] \\ &= \sqrt{v_{jl} v_{kl}} \left[c_j e^{\lambda_j t} \sqrt{\frac{v_{jl}}{v_{kl}}} + c_k e^{\lambda_k t} \sqrt{\frac{v_{kl}}{v_{jl}}} \right] \\ &= c_j v_{jl} e^{\lambda_j t} + c_k v_{kl} e^{\lambda_k t} \end{aligned} \tag{2.104}$$

when $v_{jl},\ v_{kl} > 0$. In addition, it can also be checked that

$$p_j(t - \phi_{jl}/\omega_j) z_{jl} = -c_j |v_{jl}| e^{\lambda_j t} - c_k |v_{kl}| e^{\lambda_k t} \tag{2.105}$$

when $v_{jl},\ v_{kl} < 0$, or

$$p_j(t - \phi_{jl}/\omega_j) z_{jl} = -c_j |v_{jl}| e^{\lambda_j t} + c_k v_{kl} e^{\lambda_k t} \tag{2.106}$$

when $v_{jl} < 0,\ v_{kl} > 0$, or

$$p_j(t - \phi_{jl}/\omega_j)z_{jl} = c_j v_{jl} e^{\lambda_j t} - c_k |v_{kl}| e^{\lambda_k t} \tag{2.107}$$

when $v_{jl} > 0,\ v_{kl} < 0$.

In practice, it is not necessary to work with complex time shifts because the time-shifting decoupling transformation (2.103) can be cast as a state-space transformation just like Eq. (2.48):

$$\begin{bmatrix} \mathbf{p}(t) \\ \dot{\mathbf{p}}(t) \end{bmatrix} = \begin{bmatrix} \mathbf{I} & \mathbf{I} \\ \mathbf{\Lambda}_1 & \mathbf{\Lambda}_2 \end{bmatrix} \begin{bmatrix} \mathbf{V}_1 & \mathbf{V}_2 \\ \mathbf{V}_1\mathbf{\Lambda}_1 & \mathbf{V}_2\mathbf{\Lambda}_2 \end{bmatrix}^{-1} \begin{bmatrix} \mathbf{q}(t) \\ \dot{\mathbf{q}}(t) \end{bmatrix} = \mathbf{S} \begin{bmatrix} \mathbf{q}(t) \\ \dot{\mathbf{q}}(t) \end{bmatrix}, \tag{2.108}$$

where $\mathbf{V}_2 = \widetilde{\mathbf{V}}_1$ and $\mathbf{\Lambda}_2 = \widetilde{\mathbf{\Lambda}}_1$.

2.4.2.1 Example 2.7

Consider the damped system

$$\ddot{\mathbf{q}} + \begin{bmatrix} 4 & -1 \\ -1 & 8 \end{bmatrix} \dot{\mathbf{q}} + \begin{bmatrix} 1 & 0 \\ 0 & 4 \end{bmatrix} \mathbf{q} = \mathbf{0} \tag{2.109}$$

with $\mathbf{q}(0) = [1 \quad -1]^T$ and $\dot{\mathbf{q}}(0) = [1 \quad 1]^T$. This system is non-classically damped and cannot be decoupled by modal analysis. However, it can be decoupled by phase synchronization into real single-degree-of-freedom systems. Solution of the quadratic eigenvalue problem (2.2) yields

$$\mathbf{\Lambda}_1 = \begin{bmatrix} \lambda_1 & 0 \\ 0 & \lambda_2 \end{bmatrix} = \begin{bmatrix} -7.7417 & 0 \\ 0 & -3.4142 \end{bmatrix}, \tag{2.110}$$

$$\mathbf{\Lambda}_2 = \begin{bmatrix} \lambda_3 & 0 \\ 0 & \lambda_4 \end{bmatrix} = \begin{bmatrix} -0.5858 & 0 \\ 0 & -0.2583 \end{bmatrix} \tag{2.111}$$

and normalized eigenvectors[2]

$$\mathbf{V}_1 = \begin{bmatrix} \mathbf{v}_1 & \mathbf{v}_2 \end{bmatrix} = \begin{bmatrix} -0.2485 & 0.9759 \\ 0.9620 & 0.2858 \end{bmatrix}, \tag{2.112}$$

$$\mathbf{V}_2 = \begin{bmatrix} \mathbf{v}_3 & \mathbf{v}_4 \end{bmatrix} = \begin{bmatrix} 0.6087 & -0.9034 \\ 1.0391 & 0.1167 \end{bmatrix}. \tag{2.113}$$

The decoupled system is given by

$$\ddot{\mathbf{p}} + \mathbf{D}\dot{\mathbf{p}} + \mathbf{\Omega}\mathbf{p} = \ddot{\mathbf{p}} - (\mathbf{\Lambda}_1 + \mathbf{\Lambda}_2)\dot{\mathbf{p}} + \mathbf{\Lambda}_1\mathbf{\Lambda}_2\mathbf{p} = \mathbf{0}, \tag{2.114}$$

[2]Eigenvectors have been normalized in accordance with the scheme depicted in Appendix B.

which results in

$$\begin{bmatrix}\ddot{p}_1\\ \ddot{p}_2\end{bmatrix}+\begin{bmatrix}8.3274 & 0\\ 0 & 3.6726\end{bmatrix}\begin{bmatrix}\dot{p}_1\\ \dot{p}_2\end{bmatrix}+\begin{bmatrix}4.5350 & 0\\ 0 & 0.8820\end{bmatrix}\begin{bmatrix}p_1\\ p_2\end{bmatrix}=\begin{bmatrix}0\\ 0\end{bmatrix}. \tag{2.115}$$

The initial values of the decoupled system are

$$\begin{bmatrix}p_1(0)\\ p_2(0)\end{bmatrix}=\begin{bmatrix}-0.6418\\ -2.3142\end{bmatrix},\ \begin{bmatrix}\dot{p}_1(0)\\ \dot{p}_2(0)\end{bmatrix}=\begin{bmatrix}0.5603\\ 1.8105\end{bmatrix}. \tag{2.116}$$

The response of the decoupled system is

$$\begin{bmatrix}p_1(t)\\ p_2(t)\end{bmatrix}=\begin{bmatrix}-0.0258e^{-7.7417t}-0.6161e^{-0.5858t}\\ -0.3843e^{-3.4142t}-1.9300e^{-0.2583t}\end{bmatrix}. \tag{2.117}$$

The response $\mathbf{q}(t)$ can be obtained through the configuration-space transformation

$$\mathbf{q}=\begin{bmatrix}0.6789 & -1.0573\\ 1.0454 & 0.1029\end{bmatrix}\begin{bmatrix}p_1\\ p_2\end{bmatrix}+\begin{bmatrix}0.1198 & -0.5955\\ 0.0108 & -0.0536\end{bmatrix}\begin{bmatrix}\dot{p}_1\\ \dot{p}_2\end{bmatrix} \tag{2.118}$$

resulting in

$$\begin{aligned}\begin{bmatrix}q_1(t)\\ q_2(t)\end{bmatrix}&=-0.0258\mathbf{v}_1e^{-7.7417t}-0.6161\mathbf{v}_3e^{-0.5858t}\\ &\quad-0.3843\mathbf{v}_2e^{-3.4142t}-1.9300\mathbf{v}_4e^{-0.2583t}.\end{aligned} \tag{2.119}$$

The response of the coupled and decoupled systems are illustrated in Fig. 2.5.

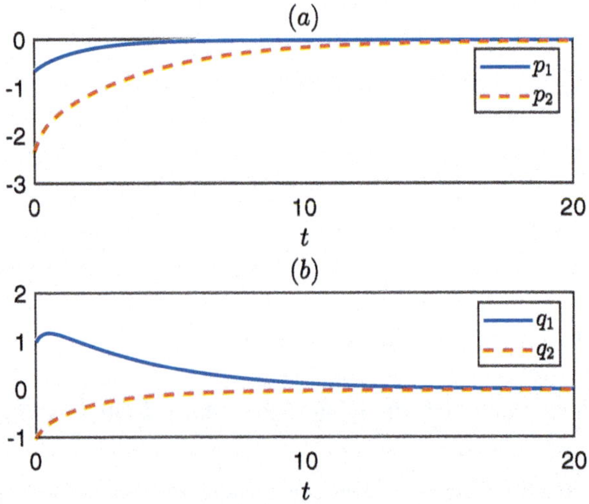

Fig. 2.5 (**a**) Response of decoupled system and (**b**) response of coupled system of Example 2.7

It is also possible to compute the coefficients of imaginary vibration. The real eigenvalues can be expressed as $\lambda_1,\ \lambda_3 = \alpha_1 \pm \omega_1$ and $\lambda_2,\ \lambda_4 = \alpha_2 \pm \omega_2$ where

$$\alpha_1 = -4.1637,\ \omega_1 = 3.5779i,\ \alpha_2 = -1.8363,\ \omega_2 = 1.5779i. \tag{2.120}$$

This means $\mathbf{p}(t)$ can be written as

$$\begin{bmatrix} p_1(t) \\ p_2(t) \end{bmatrix} = \begin{bmatrix} -0.2519e^{-4.1637t}\cos(i3.5779t + 1.5873i) \\ -1.7223e^{-1.8363t}\cos(i1.5779t + 0.8070i) \end{bmatrix}. \tag{2.121}$$

The phase angles and amplitudes can be obtained from the eigenvectors:

$$\mathbf{v}_1 = \widetilde{\mathbf{v}}_3 = \begin{bmatrix} -0.2485 \\ 0.9620 \end{bmatrix} = \begin{bmatrix} 0.3889ie^{-i(-1.5708-0.4479i)} \\ 0.9998e^{-i(-0.0386i)} \end{bmatrix}, \tag{2.122}$$

$$\mathbf{v}_2 = \widetilde{\mathbf{v}}_4 = \begin{bmatrix} 0.9759 \\ 0.2858 \end{bmatrix} = \begin{bmatrix} 0.9390ie^{-i(1.5708+0.0386i)} \\ 0.1826e^{-i(0.4479i)} \end{bmatrix}. \tag{2.123}$$

Thus the modal matrix that arises from phase synchronization is

$$\mathbf{Z} = \begin{bmatrix} \mathbf{z}_1 & \mathbf{z}_2 \end{bmatrix} = \begin{bmatrix} 0.6335 - 0.1666i & -0.8682 - 0.2283i \\ 1.0457 & 0.1084 \end{bmatrix}. \tag{2.124}$$

The decoupling transformation, as a linear time-shifting transformation, is given by

$$\begin{aligned} \mathbf{q}(t) = &\begin{bmatrix} p_1(t - \phi_{11}/\omega_1) & 0 \\ 0 & p_1(t - \phi_{12}/\omega_1) \end{bmatrix} \mathbf{z}_1 \\ &+ \begin{bmatrix} p_2(t - \phi_{21}/\omega_2) & 0 \\ 0 & p_2(t - \phi_{22}/\omega_2) \end{bmatrix} \mathbf{z}_2. \end{aligned} \tag{2.125}$$

Such transformation involves complex time-shifts. However, the overall transformation is real.

2.4.3 *Indexing and Uniqueness*

When the eigenvalues are complex, there is a unique way of pairing the eigensolutions to form the damped modes of real oscillations. Theoretically speaking, the pairing of complex conjugate eigensolutions is mandated by the constraint that $\mathbf{q}(t)$ is real and one also insists on real and physically excitable modes of vibration $\mathbf{s}_j(t)$.

This natural pairing scheme breaks down when the eigenvalues are real. The reason is that real quadratic conjugation is not unique, as opposed to complex conjugation: any two real eigensolutions may be regarded as quadratic conjugates of each other. Indeed, when all the eigenvalues are real, the eigensolutions can be paired in

$$N = \frac{\binom{2n}{2}\binom{2n-2}{2}\binom{2n-4}{2}\cdots\binom{2}{2}}{n!} = \frac{(2n)!}{2^n n!} \tag{2.126}$$

different ways to decouple the original system [28]. Many indexing schemes are possible and an indexing scheme used in [28, 30] is termed primary–secondary scheme. In it, the eigenvalues are indexed by

$$\lambda_1 < \lambda_2 < \ldots < \lambda_n < \lambda_{n+1} = \tilde{\lambda}_1 < \lambda_{n+2} = \tilde{\lambda}_2 < \ldots < \lambda_{2n} = \tilde{\lambda}_n. \tag{2.127}$$

2.4.3.1 Example 2.8

In order to illustrate the effect of indexing schemes, suppose that the eigenvalues of the system in Example 2.7 are paired in accordance with

$$\mathbf{\Lambda}_1 = \begin{bmatrix} -7.7417 & 0 \\ 0 & -3.4142 \end{bmatrix}, \ \mathbf{\Lambda}_2 = \begin{bmatrix} -0.2583 & 0 \\ 0 & -0.5858 \end{bmatrix}, \tag{2.128}$$

while the eigenvectors remain unaltered. The decoupled system would be

$$\begin{bmatrix} \ddot{p}_1 \\ \ddot{p}_2 \end{bmatrix} + \begin{bmatrix} 8 & 0 \\ 0 & 4 \end{bmatrix}\begin{bmatrix} \dot{p}_1 \\ \dot{p}_2 \end{bmatrix} + \begin{bmatrix} 2 & 0 \\ 0 & 2 \end{bmatrix}\begin{bmatrix} p_1 \\ p_2 \end{bmatrix} = \begin{bmatrix} 0 \\ 0 \end{bmatrix} \tag{2.129}$$

with

$$\begin{bmatrix} p_1(0) \\ p_2(0) \end{bmatrix} = \begin{bmatrix} -1.9557 \\ -1.0003 \end{bmatrix}, \ \begin{bmatrix} \dot{p}_1(0) \\ \dot{p}_2(0) \end{bmatrix} = \begin{bmatrix} 0.6980 \\ 1.6728 \end{bmatrix}. \tag{2.130}$$

The configuration-space transformation is now

$$\mathbf{q} = \begin{bmatrix} -0.9261 & 0.5326 \\ 0.0875 & 1.1951 \end{bmatrix}\begin{bmatrix} p_1 \\ p_2 \end{bmatrix} + \begin{bmatrix} -0.0875 & -0.1298 \\ -0.1130 & 0.2663 \end{bmatrix}\begin{bmatrix} \dot{p}_1 \\ \dot{p}_2 \end{bmatrix}. \tag{2.131}$$

2.5 Phase Synchronization of Mixed Eigensolutions

In partially oscillatory vibration, both real and complex eigenvalues of (2.2) occur simultaneously. Assume the $2n$ eigenvalues of (2.2) are distinct. Let $2n_c$ eigenvalues be complex and $2n_r = 2(n - n_c)$ be real. Separate the eigenvalues into two disjoint sets so that

$$\{\lambda\} = \{2n_c \text{ complex } \lambda\} \cup \{2n_r \text{ real } \lambda\}. \tag{2.132}$$

The complex eigenvalues occur in n_c pairs of complex conjugates, generating n_c underdamped modes. The real eigenvalues form n_r pairs of real quadratic conjugates, generating n_r overdamped modes. There is flexibility in forming each pair of real quadratic conjugates, which results in $N = (2n_r)!/2^{n_r} n_r!$ different ways to pair them. Any indexing scheme for real eigenvalues is valid, but some choices might be more convenient for computer programming when mixed eigenvalues are present [28]. To streamline computer implementations and the proceeding presentation, index the eigenvalues such that $\lambda_1, \ldots, \lambda_{n_c}$ are complex with increasing positive imaginary parts, whose complex conjugates are $\lambda_{n+1} = \bar{\lambda}_1, \ldots, \lambda_{n+n_c} = \bar{\lambda}_{n_c}$, and $\lambda_{n_c+1} < \lambda_{n_c+2} < \ldots < \lambda_n < \lambda_{n+n_c+1} < \lambda_{2n}$ are real. This means

$$\begin{aligned}\mathbf{q}(t) &= \sum_{j=1}^{2n} c_j \mathbf{v}_j e^{\lambda_j t} = \sum_{j=1}^{n} \mathbf{s}_j(t) \\ &= \sum_{j=1}^{n_c} A_j e^{\alpha_j t} \begin{bmatrix} r_{j1}\cos(\omega_j t - \phi_{j1} - \theta_j) \\ r_{j2}\cos(\omega_j t - \phi_{j2} - \theta_j) \\ \vdots \\ r_{jn}\cos(\omega_j t - \phi_{jn} - \theta_j) \end{bmatrix} \\ &+ \sum_{j=1}^{n_r} \left[c_{n_c+j} \mathbf{v}_{n_c+j} e^{\lambda_{n_c+j} t} + c_{n+n_c+j} \mathbf{v}_{n+n_c+j} e^{\lambda_{n+n_c+j} t} \right].\end{aligned} \tag{2.133}$$

A system in free vibration with mixed eigenvalues can be decoupled by dividing its eigensolutions into two groups: one associated with the real eigenvalues and another with complex eigenvalues. The complex eigensolutions are treated by complex conjugation and the real ones by real quadratic conjugation. In this way it is possible to streamline formulas for the case of mixed eigenvalues. Simultaneous phase synchronization of the damped modes of real and imaginary vibration produces a decoupled system

$$\ddot{\mathbf{p}} + \mathbf{D}\dot{\mathbf{p}} + \mathbf{\Omega}\mathbf{p} = \ddot{\mathbf{p}} - (\mathbf{\Lambda}_1 + \mathbf{\Lambda}_2)\dot{\mathbf{p}} + \mathbf{\Lambda}_1\mathbf{\Lambda}_2\mathbf{p} = \mathbf{0}, \tag{2.134}$$

where each component equation is

$$\ddot{p}_j - (\lambda_j + \lambda_{n+j})\dot{p}_j + \lambda_j \lambda_{n+j} p_j = 0s \tag{2.135}$$

with appropriate pairing of the eigenvalues. With ω representing either the imaginary part of a complex eigenvalue or the imaginary frequency associated with a pair of real eigenvalues, the time-shifting transformation is

$$\mathbf{q}(t) = \sum_{j=1}^{n} \operatorname{diag}\left[p_j(t - \phi_{j1}/\omega_j), \ldots, p_j(t - \phi_{jn}/\omega_j) \right] \mathbf{z}_j. \tag{2.136}$$

In addition, the decoupling transformation in state space is still given by the same form as in Eq. (2.48).

2.5.1 Example 2.9

Consider the damped system

$$\begin{bmatrix} 2 & 0 \\ 0 & 1 \end{bmatrix}\ddot{\mathbf{q}} + \begin{bmatrix} 2 & -2 \\ -2 & 2 \end{bmatrix}\dot{\mathbf{q}} + \begin{bmatrix} 3 & -1 \\ -1 & 1 \end{bmatrix}\mathbf{q} = \mathbf{0} \tag{2.137}$$

with

$$\mathbf{q}(0) = \begin{bmatrix} -1/2 \\ 2 \end{bmatrix}, \quad \dot{\mathbf{q}}(0) = \mathbf{0}. \tag{2.138}$$

Since $\mathbf{CM}^{-1}\mathbf{K} \neq \mathbf{KM}^{-1}\mathbf{C}$, this system is non-classically damped and cannot be decoupled by modal analysis. Phase synchronization decouples it into real single-degree-of-freedom systems. The quadratic eigenvalue problem (2.2) reveals that the eigenvalues and eigenvectors[3] can be arranged by

$$\boldsymbol{\Lambda}_1 = \begin{bmatrix} \lambda_1 & 0 \\ 0 & \lambda_2 \end{bmatrix} = \begin{bmatrix} -0.0501 + 0.7955i & 0 \\ 0 & -2.1766 \end{bmatrix}, \tag{2.139}$$

$$\boldsymbol{\Lambda}_2 = \begin{bmatrix} \lambda_3 & 0 \\ 0 & \lambda_4 \end{bmatrix} = \begin{bmatrix} -0.0501 - 0.7955i & 0 \\ 0 & -0.7232 \end{bmatrix}, \tag{2.140}$$

$$\mathbf{V}_1 = \begin{bmatrix} \mathbf{v}_1 & \mathbf{v}_2 \end{bmatrix} = \begin{bmatrix} 0.5676 + 0.0477i & -0.3664 \\ 0.6564 - 0.1706i & 0.8876 \end{bmatrix}, \tag{2.141}$$

$$\mathbf{V}_2 = \begin{bmatrix} \mathbf{v}_3 & \mathbf{v}_4 \end{bmatrix} = \begin{bmatrix} 0.5676 - 0.0477i & 0.1879 \\ 0.6564 + 0.1706i & -1.0941 \end{bmatrix}. \tag{2.142}$$

The decoupled system is given by

$$\ddot{\mathbf{p}} + \mathbf{D}\dot{\mathbf{p}} + \boldsymbol{\Omega}\mathbf{p} = \ddot{\mathbf{p}} - (\boldsymbol{\Lambda}_1 + \boldsymbol{\Lambda}_2)\dot{\mathbf{p}} + \boldsymbol{\Lambda}_1\boldsymbol{\Lambda}_2\mathbf{p} = \mathbf{0}, \tag{2.143}$$

which results in

$$\begin{bmatrix} \ddot{p}_1 \\ \ddot{p}_2 \end{bmatrix} + \begin{bmatrix} 0.1002 & 0 \\ 0 & 2.8998 \end{bmatrix}\begin{bmatrix} \dot{p}_1 \\ \dot{p}_2 \end{bmatrix} + \begin{bmatrix} 0.6353 & 0 \\ 0 & 1.5741 \end{bmatrix}\begin{bmatrix} p_1 \\ p_2 \end{bmatrix} = \begin{bmatrix} 0 \\ 0 \end{bmatrix} \tag{2.144}$$

with

$$\begin{bmatrix} p_1(0) \\ p_2(0) \end{bmatrix} = \begin{bmatrix} -0.6029 \\ -4.3348 \end{bmatrix}, \quad \begin{bmatrix} \dot{p}_1(0) \\ \dot{p}_2(0) \end{bmatrix} = \begin{bmatrix} 0.7463 \\ 4.7435 \end{bmatrix}. \tag{2.145}$$

The response is shown in Fig. 2.6.

[3]Eigenvectors have been normalized in accordance with the scheme depicted in Appendix B.

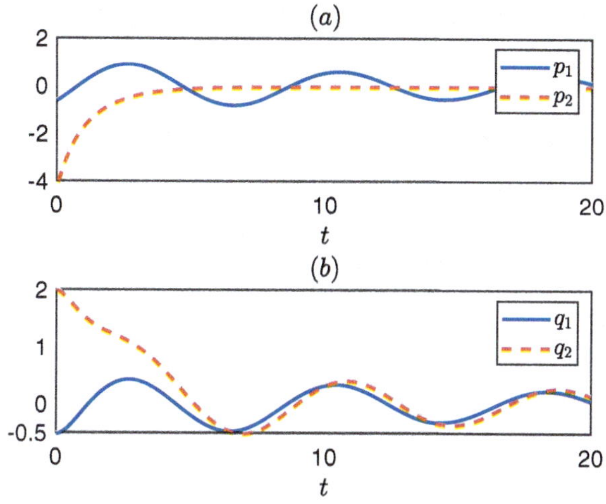

Fig. 2.6 (**a**) Response of decoupled system and (**b**) response of coupled system of Example 2.9. The decoupled system comprises one underdamped degree of freedom $p_1(t)$, formed by pairing complex conjugate eigensolutions and an overdamped degree of freedom $p_2(t)$ formed by pairing the real eigensolutions

The configuration-space decoupling transformation can be expressed as

$$\mathbf{q} = \begin{bmatrix} 0.5706 & 0.4637 \\ 0.6456 & -2.0801 \end{bmatrix} \begin{bmatrix} p_1 \\ p_2 \end{bmatrix} + \begin{bmatrix} 0.0600 & 0.3814 \\ -0.2145 & -1.3635 \end{bmatrix} \begin{bmatrix} \dot{p}_1 \\ \dot{p}_2 \end{bmatrix}. \tag{2.146}$$

2.6 Phase Synchronization of Forced Vibrations

As described earlier, the homogeneous equation (2.1) can always be decoupled by phase synchronization. Now, it needs to be determined how the excitation $\mathbf{f}(t)$ in

$$\mathbf{M}\ddot{\mathbf{q}} + \mathbf{C}\dot{\mathbf{q}} + \mathbf{K}\mathbf{q} = \mathbf{f}(t) \tag{2.147}$$

transforms under phase synchronization. Morzfeld et al. [30] postulated that the above system can be decoupled into

$$\ddot{\mathbf{p}} + \mathbf{D}\dot{\mathbf{p}} + \boldsymbol{\Omega}\mathbf{p} = \mathbf{g}(t), \tag{2.148}$$

where the coefficients are those given by the phase synchronization of the homogeneous equation: $\mathbf{D} = -(\boldsymbol{\Lambda}_1 + \boldsymbol{\Lambda}_2)$ and $\boldsymbol{\Omega} = \boldsymbol{\Lambda}_1\boldsymbol{\Lambda}_2$. Write (2.147) in first order form

$$\begin{bmatrix} \dot{\mathbf{q}}(t) \\ \ddot{\mathbf{q}}(t) \end{bmatrix} = \begin{bmatrix} \mathbf{0} & \mathbf{I} \\ -\mathbf{M}^{-1}\mathbf{K} & -\mathbf{M}^{-1}\mathbf{C} \end{bmatrix} \begin{bmatrix} \mathbf{q}(t) \\ \dot{\mathbf{q}}(t) \end{bmatrix} + \begin{bmatrix} \mathbf{0} \\ \mathbf{M}^{-1}\mathbf{f}(t) \end{bmatrix}. \tag{2.149}$$

Inspired by (2.49), define a real and invertible transformation given by

$$\begin{aligned}\begin{bmatrix}\mathbf{q}(t)\\ \dot{\mathbf{q}}(t)\end{bmatrix} &= \begin{bmatrix}\mathbf{V}_1 & \mathbf{V}_2\\ \mathbf{V}_1\boldsymbol{\Lambda}_1 & \mathbf{V}_2\boldsymbol{\Lambda}_2\end{bmatrix}\begin{bmatrix}\mathbf{I} & \mathbf{I}\\ \boldsymbol{\Lambda}_1 & \boldsymbol{\Lambda}_2\end{bmatrix}^{-1}\begin{bmatrix}\mathbf{p}_1(t)\\ \mathbf{p}_2(t)\end{bmatrix}\\ &= \mathbf{T}\begin{bmatrix}\mathbf{p}_1(t)\\ \mathbf{p}_2(t)\end{bmatrix} = \begin{bmatrix}\mathbf{T}_1 & \mathbf{T}_2\\ \mathbf{T}_3 & \mathbf{T}_4\end{bmatrix}\begin{bmatrix}\mathbf{p}_1(t)\\ \mathbf{p}_2(t)\end{bmatrix}.\end{aligned} \tag{2.150}$$

Substitution into (2.149) yields the equations

$$\mathbf{p}_2 = \dot{\mathbf{p}}_1 - \mathbf{G}_1\mathbf{f}(t), \tag{2.151}$$

$$\ddot{\mathbf{p}}_1 + \mathbf{D}\dot{\mathbf{p}}_1 + \boldsymbol{\Omega}\mathbf{p}_1 = \left(\mathbf{D} + \mathbf{I}\frac{d}{dt}\right)\mathbf{G}_1\mathbf{f}(t) + \mathbf{G}_2\mathbf{f}(t), \tag{2.152}$$

where

$$\mathbf{G}_1 = \left[\left(\mathbf{V}_1\boldsymbol{\Lambda}_1 - \mathbf{V}_2\boldsymbol{\Lambda}_2\mathbf{V}_2^{-1}\mathbf{V}_1\right)^{-1} + \left(\mathbf{V}_2\boldsymbol{\Lambda}_2 - \mathbf{V}_1\boldsymbol{\Lambda}_1\mathbf{V}_1^{-1}\mathbf{V}_2\right)^{-1}\right]\mathbf{M}^{-1} \tag{2.153}$$

and

$$\begin{aligned}\mathbf{G}_2 = \Big[&\boldsymbol{\Lambda}_1\left(\mathbf{V}_1\boldsymbol{\Lambda}_1 - \mathbf{V}_2\boldsymbol{\Lambda}_2\mathbf{V}_2^{-1}\mathbf{V}_1\right)^{-1}\\ &+ \boldsymbol{\Lambda}_2\left(\mathbf{V}_2\boldsymbol{\Lambda}_2 - \mathbf{V}_1\boldsymbol{\Lambda}_1\mathbf{V}_1^{-1}\mathbf{V}_2\right)^{-1}\Big]\mathbf{M}^{-1}.\end{aligned} \tag{2.154}$$

By inspection of Eq. (2.148), $\mathbf{p}_1(t) = \mathbf{p}(t)$ with

$$\mathbf{g}(t) = \left(\mathbf{D} + \mathbf{I}\frac{d}{dt}\right)\mathbf{G}_1\mathbf{f}(t) + \mathbf{G}_2\mathbf{f}(t). \tag{2.155}$$

The configuration-space transformation can be written as

$$\mathbf{q}(t) = \left(\mathbf{T}_1 + \mathbf{T}_2\frac{d}{dt}\right)\mathbf{p}(t) - \mathbf{T}_2\mathbf{G}_1\mathbf{f}(t). \tag{2.156}$$

This transformation is real, time-dependent, and non-linear. The non-linearity can be deduced by an examination of the form of the transformation. Transformation (2.156) has the form $\mathbf{q} = \mathbf{L}\mathbf{p} + \mathbf{c}$, which does not satisfy the superposition principle. Let $\mathbf{p}_1$ and $\mathbf{p}_2$ be two inputs and $\mathbf{q}_1$ and $\mathbf{q}_2$ the corresponding outputs:

$$\mathbf{q}_1 = \mathbf{L}\mathbf{p}_1 + \mathbf{c}, \tag{2.157}$$

$$\mathbf{q}_2 = \mathbf{L}\mathbf{p}_2 + \mathbf{c}. \tag{2.158}$$

Then, $\mathbf{q}_1 + \mathbf{q}_2 = \mathbf{L}(\mathbf{p}_1 + \mathbf{p}_2) + 2\mathbf{c}$. Thus, the input $\mathbf{p}_1 + \mathbf{p}_2$ does not produce the output $\mathbf{q}_1 + \mathbf{q}_2$, and the transformation is not linear. On the other hand, if the system is homogeneous so that $\mathbf{f}(t) = \mathbf{0}$, the decoupling transformation has the form $\mathbf{q} = \mathbf{Lp}$ and becomes linear.

The dependence of the transformation on $\mathbf{f}(t)$ was further elucidated by Morzfeld et al. [30]: if $\mathbf{f}(t) = \mathbf{0}$, viscous damping induces time shifts between the components of a mode $\mathbf{s}_j(t)$, which are accounted for in Eq. (2.136). When $\mathbf{f}(t) \neq \mathbf{0}$, the transformation must account for additional and not necessarily constant time-shifts caused by the external force, thus it is dependent on $\mathbf{f}(t)$. As a state-space transformation,

$$\begin{bmatrix} \mathbf{p}(t) \\ \dot{\mathbf{p}}(t) \end{bmatrix} = \begin{bmatrix} \mathbf{I} & \mathbf{I} \\ \mathbf{\Lambda}_1 & \mathbf{\Lambda}_2 \end{bmatrix} \begin{bmatrix} \mathbf{V}_1 & \mathbf{V}_2 \\ \mathbf{V}_1\mathbf{\Lambda}_1 & \mathbf{V}_2\mathbf{\Lambda}_2 \end{bmatrix}^{-1} \begin{bmatrix} \mathbf{q}(t) \\ \dot{\mathbf{q}}(t) \end{bmatrix} + \begin{bmatrix} \mathbf{0} \\ \mathbf{G}_1\mathbf{f}(t) \end{bmatrix}. \tag{2.159}$$

2.6.1 Example 2.10

Consider the damped system

$$\ddot{\mathbf{q}} + \begin{bmatrix} 0.7 & -0.1 \\ -0.1 & 0.2 \end{bmatrix} \dot{\mathbf{q}} + \begin{bmatrix} 2 & -1 \\ -1 & 2 \end{bmatrix} \mathbf{q} = \begin{bmatrix} f_0 \cos(\omega t) \\ 0 \end{bmatrix}. \tag{2.160}$$

This system cannot be decoupled by modal analysis. Based on Example 2.4, the decoupled system is given by

$$\ddot{\mathbf{p}} + \mathbf{D}\dot{\mathbf{p}} + \mathbf{\Omega}\mathbf{p} = \mathbf{g}(t), \tag{2.161}$$

where

$$\mathbf{D} = \begin{bmatrix} 0.3584 & 0 \\ 0 & 0.5416 \end{bmatrix}, \quad \mathbf{\Omega} = \begin{bmatrix} 1.0337 & 0 \\ 0 & 2.9022 \end{bmatrix}. \tag{2.162}$$

The responses are illustrated in Fig. 2.7.

The excitation $\mathbf{g}(t)$ of the decoupled system is given by Eq. (2.155) which results in

$$\begin{aligned} \mathbf{g}(t) = \begin{bmatrix} g_1(t) \\ g_2(t) \end{bmatrix} &= \begin{bmatrix} 0.7158 & 0.7349 \\ 0.7415 & -0.6860 \end{bmatrix} \begin{bmatrix} f_0 \cos(\omega t) \\ 0 \end{bmatrix} \\ &+ \begin{bmatrix} -0.0948 & 0.0944 \\ 0.0972 & 0.0952 \end{bmatrix} \begin{bmatrix} -\omega f_0 \sin(\omega t) \\ 0 \end{bmatrix}. \end{aligned} \tag{2.163}$$

In steady state, the modal coordinates p_1 and p_2 are periodic with the same period as $\mathbf{g}(t)$. The excitations are depicted in Fig. 2.8.

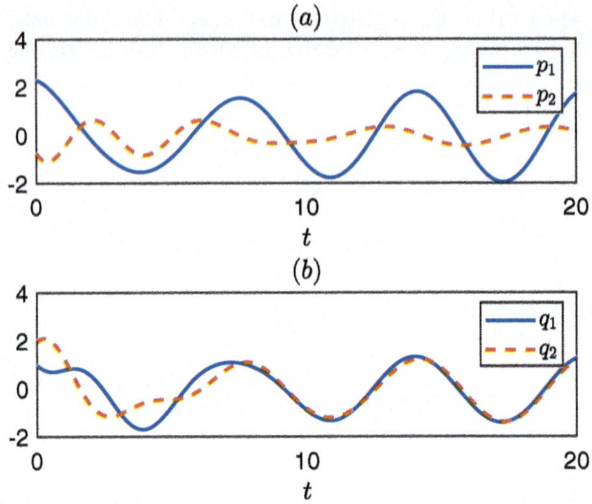

Fig. 2.7 (**a**) Response of decoupled system and (**b**) response of coupled system of Example 2.10. For this simulation $\mathbf{q}(0) = [1 \quad 2]^T$, $\dot{\mathbf{q}}(0) = [-1 \quad 1]^T$, $f_0 = 1$ and $\omega = 1$

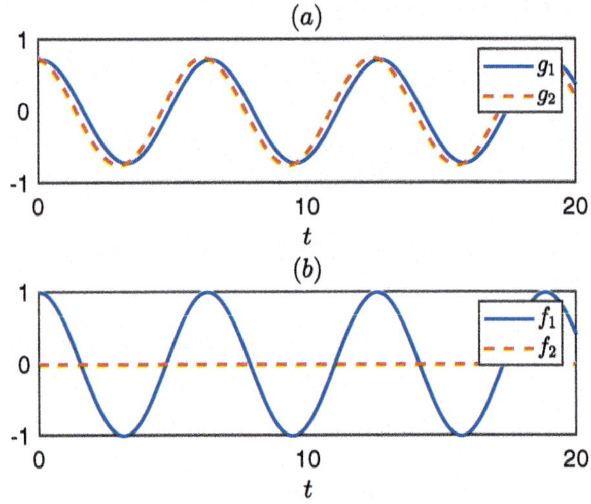

Fig. 2.8 (**a**) Excitation $\mathbf{g}(t)$ of the decoupled system and (**b**) excitation $\mathbf{f}(t)$ of coupled system of Example 2.10. For this simulation $\mathbf{q}(0) = [1 \quad 2]^T$, $\dot{\mathbf{q}}(0) = [-1 \quad 1]^T$, $f_0 = 1$ and $\omega = 1$

The configuration-space decoupling transformation can be expressed as

$$\mathbf{q}(t) = \begin{bmatrix} 0.7158 & 0.7415 \\ 0.7349 & -0.6860 \end{bmatrix} \mathbf{p}(t) + \begin{bmatrix} -0.0948 & 0.0972 \\ 0.0944 & 0.0952 \end{bmatrix} \dot{\mathbf{p}}(t) \\ - \begin{bmatrix} 0.0184 & 0.0003 \\ 0.0003 & 0.0180 \end{bmatrix} \begin{bmatrix} f_0 \cos(\omega t) \\ 0 \end{bmatrix}. \tag{2.164}$$

In the steady state, the modal coordinates p_1 and p_2 are the real part of

$$\begin{aligned} p_1(t) &= H_1(i\omega) g_1(t) \\ &= \frac{0.7158 - 0.0948\omega}{(i\omega)^2 + 0.3584(i\omega) + 1.0337} f_0 e^{i\omega t} = A_1 f_0 e^{i\omega t} \end{aligned} \tag{2.165}$$

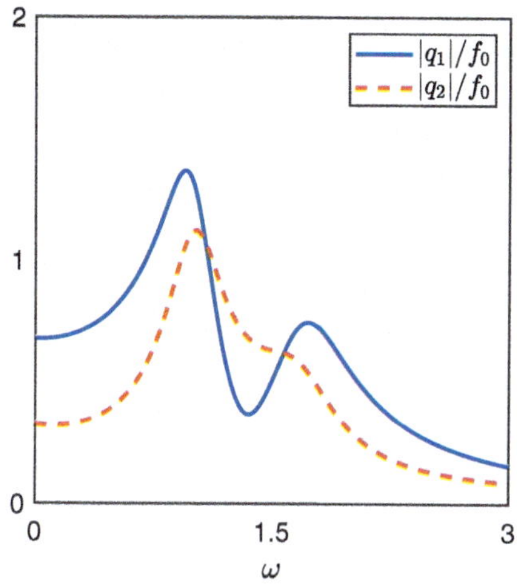

Fig. 2.9 The amplitudes $|q_1|$ and $|q_2|$ have peaks when the driving frequency ω coincides with the system natural frequencies $\omega_1 = 1$ and $\omega_2 = \sqrt{3}$. The response curves attain a relative minimum when $\omega = \sqrt{2}$. These features are similar to those of a classically damped vibration absorber

and

$$\begin{aligned} p_2(t) &= H_2(i\omega)g_2(t) \\ &= \frac{0.7415 + 0.0972\omega}{(i\omega)^2 + 0.5416(i\omega) + 2.9022} f_0 e^{i\omega t} = A_2 f_0 e^{i\omega t}, \end{aligned} \tag{2.166}$$

respectively. It follows that the frequency response in coupled coordinates are

$$\frac{|q_1(t)|}{f_0} = |0.7158A_1 + 0.7415A_2 - (i\omega)0.0948A_1 + (i\omega)0.0972A_2|, \tag{2.167}$$

$$\frac{|q_2(t)|}{f_0} = |0.7349A_1 - 0.6860A_2 + (i\omega)0.0944A_1 + (i\omega)0.0952A_2|. \tag{2.168}$$

The steady-state responses $q_1(t)$ and $q_2(t)$ are harmonic with the same frequency ω as the excitation. The frequency response curves are shown in Fig. 2.9.

2.7 Phase Synchronization as Extension of Modal Analysis

The method of phase synchronization is a direct generalization of modal analysis. To see that, consider the case system (2.1) is undamped or classically damped. In this case, the $2n$ eigenvalues occur in n pairs $\{\lambda_j,\ \lambda_{n+j}\}$, each pair sharing a common real eigenvector $\mathbf{v}_j$ so that the phase angles $\phi_{j1} = \phi_{j2} = \cdots = \phi_{jn} = 0$ or 180°. Thus, the modal vectors $\mathbf{z}_j = \mathbf{u}_j$ and Eq. (2.136) simplify to $\mathbf{q}(t) = \sum_{j=1}^{n} p_j(t)\mathbf{u}_j$, which is the modal transformation $\mathbf{q} = \mathbf{Up}$.

Alternatively, if Eq. (2.147) represents an undamped or classically damped system, $\mathbf{V}_1 = \mathbf{V}_2 = \mathbf{U}$ up to arbitrary signs in the columns of the mass-normalized modal matrix $\mathbf{U}$. Consequently,

$$\mathbf{T}_1 = \mathbf{U}, \ \mathbf{T}_2 = \mathbf{0}. \tag{2.169}$$

In this case, the configuration-space decoupling transformation represented by Eq. (2.156) reduces to the modal transformation $\mathbf{q} = \mathbf{U}\mathbf{p}$.

2.7.1 *Example 2.11*

The spectrum of the classically damped system of Example 2.1 is composed of

$$\mathbf{\Lambda}_1 = \bar{\mathbf{\Lambda}}_2 = \begin{bmatrix} -\frac{3}{8} + i\frac{\sqrt{55}}{8} & 0 \\ 0 & -\frac{5}{8} + i\frac{\sqrt{167}}{8} \end{bmatrix}, \tag{2.170}$$

$$\mathbf{V}_1 = \mathbf{V}_2 = \frac{1}{\sqrt{2}}\begin{bmatrix} 1 & 1 \\ 1 & -1 \end{bmatrix}. \tag{2.171}$$

Since $\mathbf{V}_2 = \mathbf{V}_1$, the transformation matrices in Eqs. (2.50) and (2.51) reduce to

$$\mathbf{T}_1 = \mathbf{V}_1 = \mathbf{U}, \ \mathbf{T}_2 = \mathbf{0}, \tag{2.172}$$

where $\mathbf{U}$ is the mass-normalized modal matrix. This means the decoupling transformation reduces to the modal transformation

$$\mathbf{q}(t) = \left(\mathbf{T}_1 + \mathbf{T}_2\frac{d}{dt}\right)\mathbf{p}(t) = \mathbf{U}\mathbf{p}(t) \tag{2.173}$$

and the decoupled system is

$$\ddot{\mathbf{p}} + \frac{1}{4}\begin{bmatrix} 3 & 0 \\ 0 & 5 \end{bmatrix}\dot{\mathbf{p}} + \begin{bmatrix} 1 & 0 \\ 0 & 3 \end{bmatrix}\mathbf{p} = \mathbf{0}. \tag{2.174}$$

2.8 Conclusions

In the absence of viscous damping, a linear system possesses classical normal modes, which constitute a linear coordinate transformation that decouples the undamped system. This process of decoupling the equation of motion of a damped system is the time-honored procedure termed modal analysis. A viscously damped linear system cannot be decoupled by modal analysis unless it also possesses a full set of classical normal modes, in which case the system is said to be classically damped. In general, this condition is not satisfied and thus damped linear systems cannot be decoupled by modal analysis.

Classical modal analysis is extended to decouple damped linear systems in configuration space through a procedure called phase synchronization. Similar to modal analysis, decoupling by phase synchronization possesses ample physical insight. Based upon a consideration of the physics of damping, phase angles are shifted in each non-classically damped mode of vibration so as to transform it into a classically damped mode. Unlike any methods of decoupling approximation, phase synchronization is an exact decoupling technique that accounts for all effects of coupling in full. Moreover, phase synchronization reduces to classical modal analysis under classical damping.

Chapter 3
Decoupling of Linear Systems by Phase Synchronization

The profound study of nature is the most fertile source of mathematical discoveries

—Joseph Fourier

3.1 Introduction

In Chap. 2 it was shown that the method of phase synchronization is an extension of modal analysis capable of decoupling non-classically damped systems in the real configuration space. Can the method be applied to general non-conservative systems (1.11)? How to handle repeated and defective eigenvalues? This chapter aims to answer these questions. Special attention is given in Sect. 3.2 to systems of the form (1.11) possessing distinct eigenvalues because those are of particular interest to engineers. Systems possessing repeated and defective eigenvalues are addressed in Sect. 3.3.

3.2 Linear Systems Possessing Distinct Eigenvalues

Consider the general linear system of the form

$$\mathbf{A}_2\ddot{\mathbf{q}} + \mathbf{A}_1\dot{\mathbf{q}} + \mathbf{A}_0\mathbf{q} = \mathbf{f}(t), \tag{3.1}$$

where the coefficient matrices are just real $n \times n$ matrices. Assume that the eigenvalues

$$(\lambda^2\mathbf{A}_2 + \lambda\mathbf{A}_1 + \mathbf{A}_0)\mathbf{v} = \mathbf{0} \tag{3.2}$$

are distinct. Can phase synchronization be applied to decouple this system into

$$\ddot{\mathbf{p}} + \mathbf{D}\dot{\mathbf{p}} + \boldsymbol{\Omega}\mathbf{p} = \mathbf{g}(t), \tag{3.3}$$

where $\mathbf{D} = -\text{diag}[\lambda_j + \lambda_{n+j}]$ and $\boldsymbol{\Omega} = \text{diag}[\lambda_j\lambda_{n+j}]$?

R. Gonçalves Salsa Junior, F. Ma, *Advances in the Theory of System Decoupling*, https://doi.org/10.1007/978-3-030-60846-0_3

An attentive reader would note that, in Chap. 2, symmetry and positive-definiteness of **M**, **C**, and **K** have not been used in the development of phase synchronization. More precisely, the lack of these properties would permit the real part α_j of the eigenvalues λ_j to be non-negative, with possibility of exponentially increasing modes of vibration. Therefore, the devised methodology can be used to decouple systems with non-symmetric coefficients, for example, gyroscopic and circulatory systems [30]. Concisely speaking, an implementation of phase synchronization to decouple systems with distinct eigenvalues is summarized as a series of tasks.

Task 1: Solve the Quadratic Eigenvalue Problem (3.2) **and Index the Eigensolutions**
Based on the developments in Chap. 2, complex conjugate eigensolutions must always be paired. If $2n_c$ eigenvalues are complex, the n_c complex eigenvalues with positive imaginary parts are arranged in order of increasing magnitude of their imaginary parts as the first n_c eigenvalues such that

$$S_1 = \{\lambda_1, \lambda_2, \ldots, \lambda_{n_c} : 0 < \text{Im}[\lambda_1] \leq \text{Im}[\lambda_2] \leq \ldots \leq \text{Im}[\lambda_{n_c}]\}. \tag{3.4}$$

Enumerate the remaining n_c complex eigenvalues, which are the complex conjugates with negative imaginary parts, in such a way that

$$S_3 = \{\lambda_{n+1} = \bar{\lambda}_1, \lambda_{n+2} = \bar{\lambda}_2, \ldots, \lambda_{n+n_c} = \bar{\lambda}_{n_c}\}. \tag{3.5}$$

Thus, $S_1 \cup S_3$ contains the entire set of $2n_c$ complex eigenvalues. The remaining $2n_r = 2(n - n_c)$ distinct real eigenvalues can be paired in

$$N = \frac{(2n_r)!}{2^{n_r} n_r!} \tag{3.6}$$

different ways. To streamline the presentation, the real eigenvalues are arranged in accordance with a primary–secondary pairing scheme such that

$$S_2 = \{\lambda_{n_c+1}, \lambda_{n_c+2}, \ldots, \lambda_n : \lambda_{n_c+1} < \lambda_{n_c+2} < \ldots < \lambda_n\} \tag{3.7}$$

and

$$\begin{aligned} S_4 = \{&\lambda_{n+n_c+1}, \lambda_{n+n_c+2}, \ldots, \lambda_{2n} : \\ &\lambda_n < \lambda_{n+n_c+1} < \lambda_{n+n_c+2} < \ldots < \lambda_{2n}\}. \end{aligned} \tag{3.8}$$

Thus, $S_2 \cup S_4$ contains the entire set of $2n_r$ real eigenvalues under the constraint that $\sup(S_2) < \inf(S_4)$. The $2n$ eigenvalues are partitioned into four disjoint subsets. This indexing scheme is illustrated in Fig. 3.1. A different indexing scheme for the real eigensolutions may be used.

Task 2: Construct the Decoupled Form (3.3) **Using the Eigenvalues and Eigenvectors of** (3.2)
Using the indexed eigensolutions, assemble the following $n \times n$ diagonal matrices

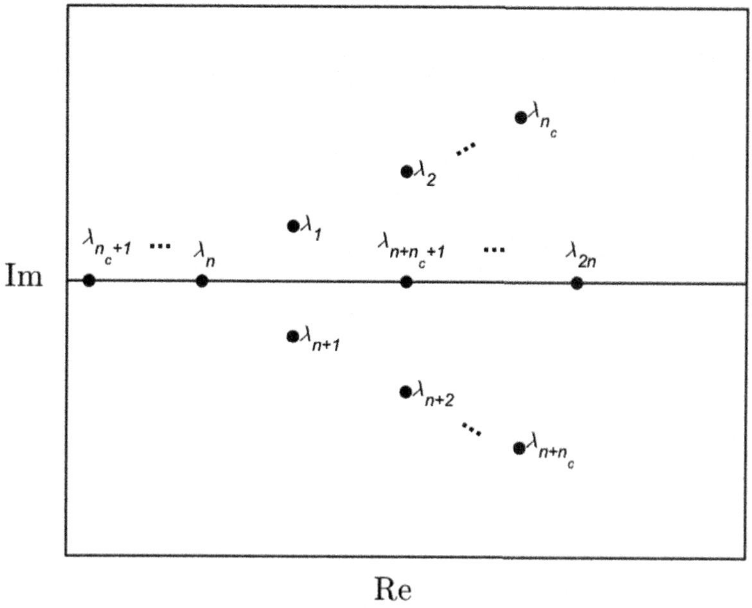

Fig. 3.1 Indexing scheme proposed in Task 1. Different indexing schemes for real eigensolutions may be used

$$\boldsymbol{\Lambda}_1 = \begin{bmatrix} \lambda_1 & & \\ & \ddots & \\ & & \lambda_n \end{bmatrix}, \quad \boldsymbol{\Lambda}_2 = \begin{bmatrix} \lambda_{n+1} & & \\ & \ddots & \\ & & \lambda_{2n} \end{bmatrix} \tag{3.9}$$

and

$$\mathbf{V}_1 = \begin{bmatrix} \mathbf{v}_1 & \mathbf{v}_2 & \dots & \mathbf{v}_n \end{bmatrix}, \quad \mathbf{V}_2 = \begin{bmatrix} \mathbf{v}_{n+1} & \mathbf{v}_{n+2} & \dots & \mathbf{v}_{2n} \end{bmatrix}. \tag{3.10}$$

The real and diagonal coefficients of equation (3.3) are given by

$$\mathbf{D} = -\left(\boldsymbol{\Lambda}_1 + \boldsymbol{\Lambda}_2\right), \quad \boldsymbol{\Omega} = \boldsymbol{\Lambda}_1 \boldsymbol{\Lambda}_2. \tag{3.11}$$

The excitation $\mathbf{g}(t)$ of (3.3) is given in terms of $\mathbf{f}(t)$ by

$$\mathbf{g}(t) = \left(\mathbf{D} + \mathbf{I}\frac{d}{dt}\right) \mathbf{G}_1 \mathbf{f}(t) + \mathbf{G}_2 \mathbf{f}(t), \tag{3.12}$$

where $\mathbf{G}_1$ and $\mathbf{G}_2$ are real $n \times n$ matrices computed in accordance with

$$\mathbf{G}_1 = \left[\left(\mathbf{V}_1 \boldsymbol{\Lambda}_1 - \mathbf{V}_2 \boldsymbol{\Lambda}_2 \mathbf{V}_2^{-1} \mathbf{V}_1 \right)^{-1} + \left(\mathbf{V}_2 \boldsymbol{\Lambda}_2 - \mathbf{V}_1 \boldsymbol{\Lambda}_1 \mathbf{V}_1^{-1} \mathbf{V}_2 \right)^{-1} \right] \boldsymbol{\Lambda}_2^{-1} \tag{3.13}$$

and

$$\mathbf{G}_2 = \Big[\mathbf{\Lambda}_1\left(\mathbf{V}_1\mathbf{\Lambda}_1 - \mathbf{V}_2\mathbf{\Lambda}_2\mathbf{V}_2^{-1}\mathbf{V}_1\right)^{-1} + \mathbf{\Lambda}_2\left(\mathbf{V}_2\mathbf{\Lambda}_2 - \mathbf{V}_1\mathbf{\Lambda}_1\mathbf{V}_1^{-1}\mathbf{V}_2\right)^{-1}\Big]\mathbf{\Lambda}_2^{-1}. \tag{3.14}$$

Task 3: Construct the Real Decoupling Transformations in the Configuration and State Spaces
Assemble the following real $n \times n$ matrices:

$$\mathbf{T}_1 = (\mathbf{V}_1\mathbf{\Lambda}_2 - \mathbf{V}_2\mathbf{\Lambda}_1)(\mathbf{\Lambda}_2 - \mathbf{\Lambda}_1)^{-1}, \tag{3.15}$$

$$\mathbf{T}_2 = (\mathbf{V}_2 - \mathbf{V}_1)(\mathbf{\Lambda}_2 - \mathbf{\Lambda}_1)^{-1}. \tag{3.16}$$

The configuration-space decoupling transformation can be expressed as

$$\mathbf{q}(t) = \left(\mathbf{T}_1 + \mathbf{T}_2\frac{d}{dt}\right)\mathbf{p}(t) - \mathbf{T}_2\mathbf{G}_1\mathbf{f}(t). \tag{3.17}$$

When cast in the state space, the decoupling transformation takes the form

$$\begin{bmatrix}\mathbf{p}(t)\\ \dot{\mathbf{p}}(t)\end{bmatrix} = \mathbf{S}\begin{bmatrix}\mathbf{q}(t)\\ \dot{\mathbf{q}}(t)\end{bmatrix} + \begin{bmatrix}\mathbf{0}\\ \mathbf{G}_1\mathbf{f}(t)\end{bmatrix}, \tag{3.18}$$

where the $2n \times 2n$ real and invertible matrix $\mathbf{S}$ is given by

$$\mathbf{S} = \begin{bmatrix}\mathbf{S}_1 & \mathbf{S}_2\\ \mathbf{S}_3 & \mathbf{S}_4\end{bmatrix} = \begin{bmatrix}\mathbf{T}_1 & \mathbf{T}_2\\ \mathbf{T}_3 & \mathbf{T}_4\end{bmatrix}^{-1} = \begin{bmatrix}\mathbf{I} & \mathbf{I}\\ \mathbf{\Lambda}_1 & \mathbf{\Lambda}_2\end{bmatrix}\begin{bmatrix}\mathbf{V}_1 & \mathbf{V}_2\\ \mathbf{V}_1\mathbf{\Lambda}_1 & \mathbf{V}_2\mathbf{\Lambda}_2\end{bmatrix}^{-1}. \tag{3.19}$$

The upper-half of equation (3.18) yields a configuration-space mapping from $\mathbf{q}$ to $\mathbf{p}$ that is an inverse of equation (3.17). When $t = 0$, Eq. (3.18) generates the initial values $\mathbf{p}(0)$ and $\dot{\mathbf{p}}(0)$ of Eq. (3.3). The decoupling transformations in both the configuration and state spaces are non-linear for non-homogeneous systems and linear for homogeneous systems.

3.2.1 *Decoupling Algorithm*

Tasks 1–3 offer an algorithm for decoupling a system by phase synchronization, which is illustrated schematically in Fig. 3.2. The algorithm provided by the flowchart was coded in two MATLAB files that are supplemented with this book. Appendix B illustrates how to generate desired outputs for a given linear system using the program files.

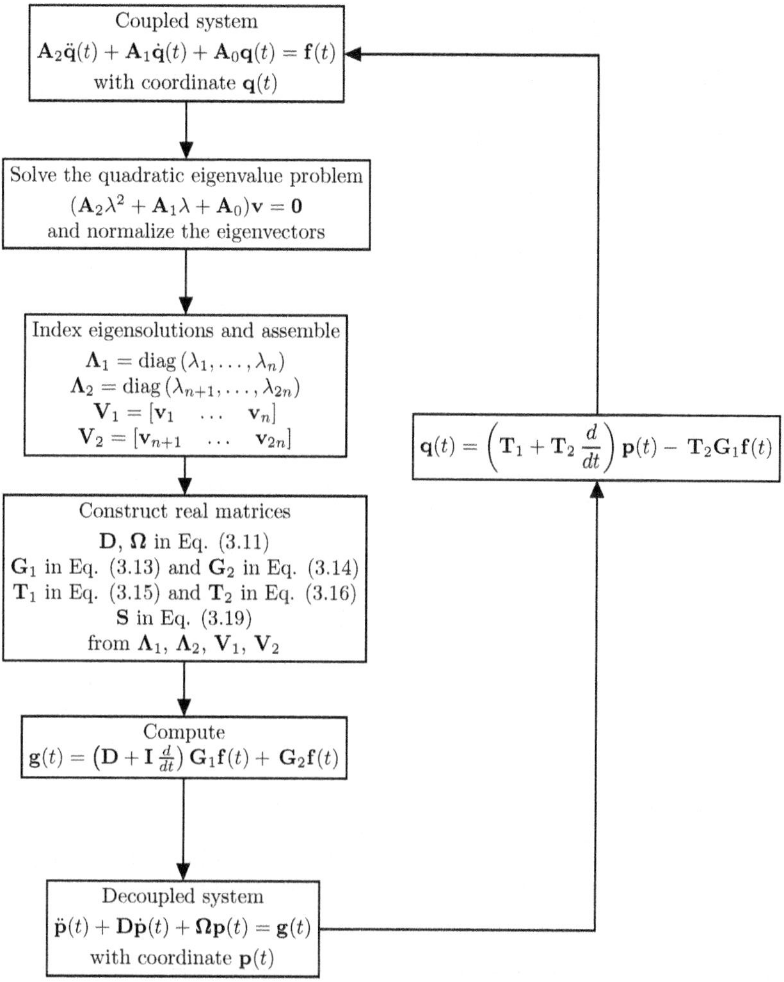

Fig. 3.2 Algorithm for decoupling a linear system with distinct eigenvalues. In the event that such system is classically damped, $\mathbf{T}_1 = \mathbf{U}$, $\mathbf{T}_2 = \mathbf{0}$ and the decoupling transformation is the modal transformation $\mathbf{q} = \mathbf{U}\mathbf{p}$. A flowchart for decoupling via modal analysis is shown in Fig. 1.6. Appendix B illustrates how this algorithm may be implemented in MATLAB

3.2.1.1 Example 3.1

Consider the non-symmetric system

$$\ddot{\mathbf{q}} + \begin{bmatrix} 1 & 1 \\ -2 & 3 \end{bmatrix}\dot{\mathbf{q}} + \begin{bmatrix} -7 & 4 \\ -8 & 1 \end{bmatrix}\mathbf{q} = \mathbf{0} \tag{3.20}$$

with $\mathbf{q}(0) = [1 \quad 2]^T$ and $\dot{\mathbf{q}}(0) = [-1 \quad 1]^T$. In Example 1.10 it was shown that this system can be decoupled in configuration space by a similarity transformation. However, the decoupled system and the decoupling transformation found therein are complex, greatly diminishing physical insight. Using phase synchronization, this system can be decoupled into real single-degree-of-freedom systems. Solution of the quadratic eigenvalue problem (3.2) reveals that[1]

$$\mathbf{\Lambda}_1 = \bar{\mathbf{\Lambda}}_2 = \begin{bmatrix} -3.0679 + 0.2254i & 0 \\ 0 & 1.0679 + 1.2254i \end{bmatrix}, \tag{3.21}$$

$$\mathbf{V}_1 = \overline{\mathbf{V}}_2 = \begin{bmatrix} -0.0836 + 0.1975i & 0.1950 + 0.4605i \\ -0.2811 + 0.1139i & 0.6555 + 0.2655i \end{bmatrix}. \tag{3.22}$$

The eigenvectors are normalized. The decoupled system, as given by (3.3), is

$$\ddot{\mathbf{p}} + \begin{bmatrix} 6.1358 & 0 \\ 0 & -2.1358 \end{bmatrix} \dot{\mathbf{p}} + \begin{bmatrix} 9.4627 & 0 \\ 0 & 2.6419 \end{bmatrix} \mathbf{p} = \mathbf{0} \tag{3.23}$$

with

$$\mathbf{p}(0) = \begin{bmatrix} -0.3679 \\ 2.4633 \end{bmatrix}, \ \dot{\mathbf{p}}(0) = \begin{bmatrix} 1.3260 \\ 3.4719 \end{bmatrix}. \tag{3.24}$$

The configuration-space decoupling transformation can be expressed as

$$\begin{aligned} \mathbf{q}(t) &= \left(\mathbf{T}_1 + \mathbf{T}_2 \frac{d}{dt} \right) \mathbf{p}(t) \\ &= \begin{bmatrix} 2.6047 & -0.2063 \\ 1.2690 & 0.4241 \end{bmatrix} \mathbf{p}(t) + \begin{bmatrix} 0.8763 & 0.3758 \\ 0.5053 & 0.2167 \end{bmatrix} \dot{\mathbf{p}}(t). \end{aligned} \tag{3.25}$$

The response is shown in Fig. 3.3.

Alternatively, it is possible to compute the time-shifting decoupling transformation. In polar form, the eigenvectors can be written as

$$\mathbf{v}_1 = \overline{\mathbf{v}}_3 = \begin{bmatrix} -0.2145 e^{-i67.0525^\circ} \\ -0.3033 e^{-i22.0525^\circ} \end{bmatrix}, \tag{3.26}$$

$$\mathbf{v}_2 = \overline{\mathbf{v}}_4 = \begin{bmatrix} -0.5001 e^{-i112.9475^\circ} \\ -0.7072 e^{-i157.9475^\circ} \end{bmatrix}. \tag{3.27}$$

[1]Numerical results obtained using the supplied MATLAB files `MAIN.m` and `SystemDecoupling.m`. See Appendix B for instructions on how to use the program.

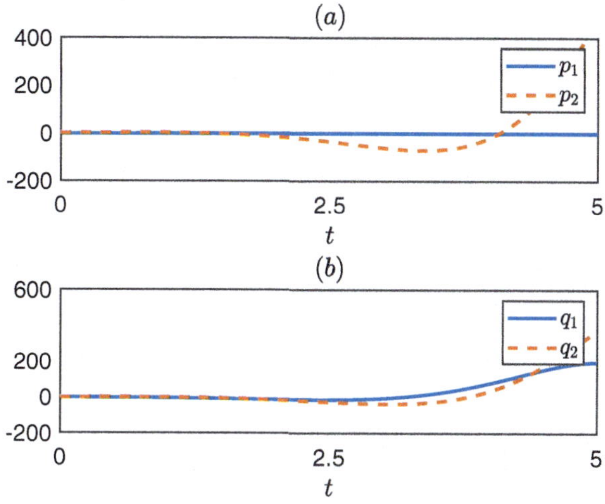

Fig. 3.3 (**a**) Response of the decoupled system and (**b**) response of coupled system of Example 3.1. Since λ_2 has positive real part, the solution is increasing exponentially

The modes of vibration are

$$\mathbf{s}_1(t) = A_1 e^{-3.0679t} \begin{bmatrix} -0.2145\cos(0.2254t - 67.0525^\circ - \theta_1) \\ -0.3033\cos(0.2254t - 22.0525^\circ - \theta_1) \end{bmatrix}, \tag{3.28}$$

$$\mathbf{s}_2(t) = A_2 e^{1.0679t} \begin{bmatrix} -0.5001\cos(1.2254t - 112.9475^\circ - \theta_2) \\ -0.7072\cos(1.2254t - 157.9475^\circ - \theta_2) \end{bmatrix}. \tag{3.29}$$

There is a constant phase difference of 45° between the two elements of $\mathbf{s}_1(t)$ as well as of $\mathbf{s}_2(t)$. The modal matrix that arises from phase synchronization is

$$\mathbf{Z} = \begin{bmatrix} 0 & -2.7869 \\ -0.0016 & -7.8144 \end{bmatrix}. \tag{3.30}$$

The decoupling transformation as a time-shifting transformation is

$$\begin{aligned} \mathbf{q}(t) &= \begin{bmatrix} p_1(t - \phi_{11}/\omega_1) & 0 \\ 0 & p_1(t - \phi_{12}/\omega_1) \end{bmatrix} \mathbf{z}_1 \\ &\quad + \begin{bmatrix} p_2(t - \phi_{21}/\omega_2) & 0 \\ 0 & p_1(t - \phi_{22}/\omega_2) \end{bmatrix} \mathbf{z}_2 \\ &= \begin{bmatrix} p_1(t - 5.1926) & 0 \\ 0 & p_1(t - 1.7078) \end{bmatrix} \begin{bmatrix} 0 \\ -0.0016 \end{bmatrix} \\ &\quad + \begin{bmatrix} p_2(t - 1.6087) & 0 \\ 0 & p_1(t - 2.2497) \end{bmatrix} \begin{bmatrix} -2.7869 \\ -7.8144 \end{bmatrix}. \end{aligned} \tag{3.31}$$

3.2.1.2 Example 3.2

Consider the non-symmetric system

$$\begin{bmatrix} 2 & -1 \\ 1 & 0 \end{bmatrix} \ddot{\mathbf{q}} + \begin{bmatrix} 1 & 0 \\ 1 & 0 \end{bmatrix} \dot{\mathbf{q}} + \begin{bmatrix} 1 & -1 \\ 0 & 0 \end{bmatrix} \mathbf{q} = \mathbf{0} \tag{3.32}$$

with $\mathbf{q}(0) = [1 \quad 2]^T$ and $\dot{\mathbf{q}}(0) = [-1 \quad 1]^T$. In Example 1.11 it was shown that such system can be converted into upper triangular form. Moreover, it was shown in Example 1.12 that this system can even be diagonalized by a common equivalence transformation, resulting in a decoupled system with singular acceleration coefficient.

Solution of the quadratic eigenvalue problem (3.2) reveals that

$$\mathbf{\Lambda}_1 = \begin{bmatrix} i & 0 \\ 0 & -1 \end{bmatrix}, \quad \mathbf{\Lambda}_2 = \begin{bmatrix} -i & 0 \\ 0 & 0 \end{bmatrix}, \tag{3.33}$$

$$\mathbf{V}_1 = \begin{bmatrix} 0 & 0.5 \\ -0.7071 & 0.5 \end{bmatrix}, \quad \mathbf{V}_2 = \begin{bmatrix} 0 & 0.7071 \\ -0.7071 & 0.7071 \end{bmatrix}. \tag{3.34}$$

The eigenvectors are normalized in accordance with the scheme depicted in Appendix B. The decoupled system generated by phase synchronization is

$$\ddot{\mathbf{p}} + \begin{bmatrix} 0 & 0 \\ 0 & 1 \end{bmatrix} \dot{\mathbf{p}} + \begin{bmatrix} 1 & 0 \\ 0 & 0 \end{bmatrix} \mathbf{p} = \mathbf{0} \tag{3.35}$$

with

$$\mathbf{p}(0) = \begin{bmatrix} -1.4142 \\ 2 \end{bmatrix}, \quad \dot{\mathbf{p}}(0) = \begin{bmatrix} -2.8284 \\ -2 \end{bmatrix}. \tag{3.36}$$

The configuration-space decoupling transformation can be expressed as

$$\begin{aligned} \mathbf{q}(t) &= \left(\mathbf{T}_1 + \mathbf{T}_2 \frac{d}{dt} \right) \mathbf{p}(t) \\ &= \begin{bmatrix} 0 & 0.7071 \\ -0.7071 & 0.7071 \end{bmatrix} \mathbf{p}(t) + \begin{bmatrix} 0 & 0.2071 \\ 0 & 0.2071 \end{bmatrix} \dot{\mathbf{p}}(t). \end{aligned} \tag{3.37}$$

The response is shown in Fig. 3.4.

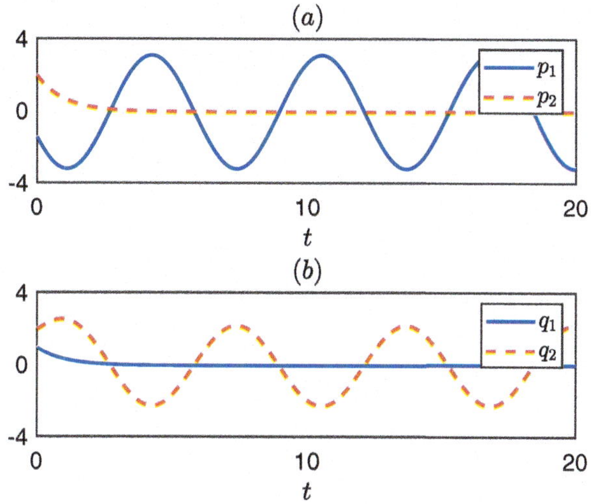

Fig. 3.4 (**a**) Response of the decoupled system and (**b**) response of coupled system of Example 3.2

3.2.2 *Phase Synchronization and Structure-Preserving Transformations*

Two systems with identical eigenvalues and multiplicities are termed strictly isospectral [30, 32]. Since the property of being strictly isospectral is reflexive, transitive, and symmetric, strictly isospectral systems generate an equivalence class [80]. It is easy to verify that system (3.1) and the decoupled system (3.3) are strictly isospectral regardless of the pairing of real eigensolutions during phase synchronization. Indeed, every system within the equivalence class can be generated by suitably pairing the real eigensolutions. Thus, phase synchronization generates the path to all real and diagonal systems within the equivalence class of systems strictly isospectral to (3.1) [30]. Alternatively, observe that the homogeneous equations in state space for the coupled and decoupled systems are, respectively,

$$\begin{bmatrix}\dot{\mathbf{q}}\\ \ddot{\mathbf{q}}\end{bmatrix} = \begin{bmatrix}\mathbf{0} & \mathbf{I}\\ -\mathbf{A}_2^{-1}\mathbf{A}_0 & -\mathbf{A}_2^{-1}\mathbf{A}_1\end{bmatrix}\begin{bmatrix}\mathbf{q}\\ \dot{\mathbf{q}}\end{bmatrix}, \tag{3.38}$$

$$\begin{bmatrix}\dot{\mathbf{p}}\\ \ddot{\mathbf{p}}\end{bmatrix} = \begin{bmatrix}\mathbf{0} & \mathbf{I}\\ -\boldsymbol{\Omega} & -\mathbf{D}\end{bmatrix}\begin{bmatrix}\mathbf{p}\\ \dot{\mathbf{p}}\end{bmatrix}. \tag{3.39}$$

Isospectrality implies that state-space coefficient matrices have the same eigenvalues. In addition, each is diagonalizable because the eigenvalues are assumed distinct. Recall that two diagonalizable matrices are isospectral if and only if they are similar. It can be checked by direct manipulations that

$$\begin{bmatrix}\mathbf{0} & \mathbf{I}\\ -\boldsymbol{\Omega} & -\mathbf{D}\end{bmatrix} = \mathbf{T}^{-1}\begin{bmatrix}\mathbf{0} & \mathbf{I}\\ -\mathbf{A}_2^{-1}\mathbf{A}_0 & -\mathbf{A}_2^{-1}\mathbf{A}_1\end{bmatrix}\mathbf{T} = \mathbf{T}^{-1}\mathbf{A}\mathbf{T}, \tag{3.40}$$

where $\mathbf{T}^{-1} = \mathbf{S}$ is given by Eq. (3.19) and $\mathbf{A}$ is the state companion matrix. Therefore, phase synchronization is a structure-preserving transformation (SPT).

Diagonalizing SPTs are strictly isospectral transformations concerned with finding real invertible $2n \times 2n$ matrices $\mathbf{U}_R$, $\mathbf{U}_L$ such that the linear matrix pencil [32, 34–37]

$$L(\lambda) = \begin{bmatrix} \mathbf{A}_1 & \mathbf{A}_2 \\ \mathbf{A}_2 & \mathbf{0} \end{bmatrix} \lambda + \begin{bmatrix} \mathbf{A}_0 & \mathbf{0} \\ \mathbf{0} & -\mathbf{A}_2 \end{bmatrix} \tag{3.41}$$

is transformed into

$$\mathbf{U}_L^T L(\lambda) \mathbf{U}_R = \begin{bmatrix} \mathbf{B}_1 & \mathbf{B}_2 \\ \mathbf{B}_2 & \mathbf{0} \end{bmatrix} \lambda + \begin{bmatrix} \mathbf{B}_0 & \mathbf{0} \\ \mathbf{0} & -\mathbf{B}_2 \end{bmatrix}, \tag{3.42}$$

i.e., the block structure has been preserved and the coefficients $\mathbf{B}_2$, $\mathbf{B}_1$, and $\mathbf{B}_0$ are real and diagonal. If such procedure is possible, the associated second-order decoupled equation

$$\mathbf{B}_2 \ddot{\mathbf{p}} + \mathbf{B}_1 \dot{\mathbf{p}} + \mathbf{B}_0 \mathbf{p} = \mathbf{0} \tag{3.43}$$

is obtained.

The decoupled system (3.3) obtained by phase synchronization is the real and diagonal system isospectral to (3.1). Thus, a diagonalizing SPT must generate the same decoupled system as the one produced via phase synchronization. Indeed, if the eigenvalues of Eq. (3.2) are distinct, $\mathbf{U}_R = \mathbf{U}_L = \mathbf{T}$ and $\mathbf{B}_1 = \mathbf{D}$, and $\mathbf{B}_0 = \boldsymbol{\Omega}$, provided we choose $\mathbf{B}_2 = \mathbf{I}$, which can be done without loss of generality. Thus, a diagonalizing SPT leads to the same decoupled system as obtained by phase synchronization.

3.2.2.1 Example 3.3

Consider the non-symmetric system of Example 3.2. It can be shown that

$$\mathbf{S} = \begin{bmatrix} \mathbf{I} & \mathbf{I} \\ \boldsymbol{\Lambda}_1 & \boldsymbol{\Lambda}_2 \end{bmatrix} \begin{bmatrix} \mathbf{V}_1 & \mathbf{V}_2 \\ \mathbf{V}_1\boldsymbol{\Lambda}_1 & \mathbf{V}_2\boldsymbol{\Lambda}_2 \end{bmatrix}^{-1}$$

$$= \begin{bmatrix} 1.4142 & -1.4142 & 0 & 0 \\ 1.4142 & 0 & -0.5858 & 0 \\ 0 & 0 & 1.4142 & -1.4142 \\ 0 & 0 & 2 & 0 \end{bmatrix}. \tag{3.44}$$

It can be checked that

$$\mathbf{S}\begin{bmatrix} \mathbf{0} & \mathbf{I} \\ -\mathbf{A}_2^{-1}\mathbf{A}_0 & -\mathbf{A}_2^{-1}\mathbf{A}_1 \end{bmatrix}\mathbf{S}^{-1} = \begin{bmatrix} 0 & 0 & 1 & 0 \\ 0 & 0 & 0 & 1 \\ -1 & 0 & 0 & 0 \\ 0 & 0 & 0 & -1 \end{bmatrix} = \begin{bmatrix} \mathbf{0} & \mathbf{I} \\ -\boldsymbol{\Omega} & -\mathbf{D} \end{bmatrix}. \tag{3.45}$$

Thus $\mathbf{SAS}^{-1}$ is structure-preserving.

3.3 Linear Systems Possessing Repeated and Defective Eigenvalues

It is now necessary to make generalizations to address linear systems not previously considered, such as systems possessing repeated eigenvalues or defective/degenerate systems (see Sect. 1.2.3).

3.3.1 *Phase Synchronization of Homogeneous Systems*

The solution of

$$\mathbf{A}_2\ddot{\mathbf{q}} + \mathbf{A}_1\dot{\mathbf{q}} + \mathbf{A}_0\mathbf{q} = \mathbf{0} \tag{3.46}$$

has a general form

$$\mathbf{q}(t) = \mathbf{V}_q e^{\mathbf{J}_q t}\mathbf{c} \tag{3.47}$$

regardless of the nature of its spectrum. Recall that $\mathbf{J}_q$ is an order $2n$ Jordan matrix with the eigenvalues on the diagonal, $\mathbf{V}_q$ is an $n \times 2n$ matrix of the associated eigenvectors and generalized eigenvectors, and $\mathbf{c}$ is a $2n$-dimensional column vector of coefficients determined from the initial conditions $\mathbf{q}(0)$ and $\dot{\mathbf{q}}(0)$. Eq. (3.47) can be recast as a state equation:

$$\begin{bmatrix} \mathbf{q}(t) \\ \dot{\mathbf{q}}(t) \end{bmatrix} = \begin{bmatrix} \mathbf{V}_q \\ \mathbf{V}_q\mathbf{J}_q \end{bmatrix} e^{\mathbf{J}_q t}\mathbf{c}. \tag{3.48}$$

Assume that system (3.46) can be decoupled into

$$\ddot{\mathbf{p}} + \mathbf{D}\dot{\mathbf{p}} + \boldsymbol{\Omega}\mathbf{p} = \mathbf{0}, \tag{3.49}$$

where $\mathbf{D} = -\text{diag}[\lambda_1 + \lambda_{n+1}, \ldots, \lambda_n + \lambda_{2n}]$ and $\boldsymbol{\Omega} = \text{diag}[\lambda_1\lambda_{n+1}, \ldots, \lambda_n\lambda_{2n}]$ with appropriate pairing of the eigensolutions λ_j, λ_{n+j} $(j = 1, 2, \ldots, n)$.

Phase synchronization for systems possessing distinct eigenvalues does not disturb the eigensolution coefficients $\mathbf{c}$. In this light, construct a transformation that decouples (3.46) into (3.49) while also preserving these coefficients. Consequently, the modal response $\mathbf{p}(t)$ of (3.49) may be cast as

$$\mathbf{p}(t) = \mathbf{V}_p e^{\mathbf{J}_p t}\mathbf{c}, \tag{3.50}$$

where $\mathbf{J}_p$ is a Jordan matrix of order $2n$ for system (3.49). It consists of the system eigenvalues on the diagonal entries, usually a modified form of $\mathbf{J}_q$ whose structure imposes the eigenvalue pairing scheme for decoupling. This means that the algebraic multiplicities of the eigenvalues are preserved in this decoupling procedure. If $\mathbf{J}_p = \mathbf{J}_q$, then decoupling is achieved by a strictly isospectral transformation. However, if one insists that geometric multiplicities be preserved, then a defective system may not be fully decoupled [81]. If the main goal is to decouple defective systems at any cost, then geometric multiplicities may not be preserved [31]. The conditions under which strictly isospectral decoupling is permissible are presented in Sect. 3.3.3.

Matrix $\mathbf{V}_p$ contains the eigenvectors of the decoupled system (3.49), the ordering of each depends on the chosen pairing scheme specified in $\mathbf{J}_p$. In the case two distinct eigenvalues λ_j and λ_{n+j} are to be paired, the corresponding columns in $\mathbf{V}_p$ are equal to $\mathbf{e}_j$, an n-dimensional column vector of zeros except the jth row entry is equal to one. For damped systems, these would correspond to underdamped or overdamped degrees of freedom. In the case two repeated real eigenvalues $\lambda_j = \lambda_{n+j}$ are paired, corresponding columns in $\mathbf{V}_p$ are $\mathbf{e}_j$ and $\mathbf{0}$, because it is always possible to choose $\mathbf{0}$ as the generalized eigenvector of a defective eigenvalue [32]. This would correspond to a critically damped degree of freedom for a damped system.

As a state equation, $\mathbf{p}(t)$ and $\dot{\mathbf{p}}(t)$ also satisfy

$$\begin{bmatrix} \mathbf{p}(t) \\ \dot{\mathbf{p}}(t) \end{bmatrix} = \begin{bmatrix} \mathbf{V}_p \\ \mathbf{V}_p\mathbf{J}_p \end{bmatrix} e^{\mathbf{J}_p t}\mathbf{c}. \tag{3.51}$$

Combination of Eqs. (3.48) and (3.51) yields

$$\begin{bmatrix} \mathbf{q}(t) \\ \dot{\mathbf{q}}(t) \end{bmatrix} = \begin{bmatrix} \mathbf{V}_q \\ \mathbf{V}_q\mathbf{J}_q \end{bmatrix} e^{\mathbf{J}_q t} e^{-\mathbf{J}_p t} \begin{bmatrix} \mathbf{V}_p \\ \mathbf{V}_p\mathbf{J}_p \end{bmatrix}^{-1} \begin{bmatrix} \mathbf{p}(t) \\ \dot{\mathbf{p}}(t) \end{bmatrix} = \mathbf{T}(t) \begin{bmatrix} \mathbf{p}(t) \\ \dot{\mathbf{p}}(t) \end{bmatrix}. \tag{3.52}$$

The invertibility of $\mathbf{T}$ is assured because $\mathbf{J}_q$ and $\mathbf{V}_q$ constitute a Jordan pair and the eigenvalues are paired to preserve the second-order structure of the equation of motion (3.46) [31]. The corresponding configuration-space decoupling transformation is obtained by extracting the upper-half of (3.52):

$$\mathbf{q}(t) = \left[\mathbf{T}_1(t) + \mathbf{T}_2(t)\frac{d}{dt}\right]\mathbf{p}(t). \tag{3.53}$$

The initial conditions are then connected by

$$\begin{bmatrix} \mathbf{p}(0) \\ \dot{\mathbf{p}}(0) \end{bmatrix} = \begin{bmatrix} \mathbf{V}_p \\ \mathbf{V}_p\mathbf{J}_p \end{bmatrix} \begin{bmatrix} \mathbf{V}_q \\ \mathbf{V}_q\mathbf{J}_q \end{bmatrix}^{-1} \begin{bmatrix} \mathbf{q}(0) \\ \dot{\mathbf{q}}(0) \end{bmatrix}. \tag{3.54}$$

Decoupling a defective system is less systematic than in the non-defective case, as the process of constructing the coefficient matrices $\mathbf{D}$ and $\boldsymbol{\Omega}$ of Eq. (3.49) varies with the number of real eigenvalues and with the geometric multiplicities of the defective eigenvalues. Moreover, it is generally not possible to write the time-varying transformation matrix $\mathbf{T}(t)$ in Eq. (3.52) in a simplified and descriptive form, as exemplified by Eq. (3.19) when the system is non-defective.

Finally, it is straightforward to show that the state companion matrices in Eqs. (3.38) and (3.39) are related by [31]

$$\begin{bmatrix} \mathbf{0} & \mathbf{I} \\ -\boldsymbol{\Omega} & -\mathbf{D} \end{bmatrix} = \mathbf{T}^{-1}(t) \begin{bmatrix} \mathbf{0} & \mathbf{I} \\ -\mathbf{A}_2^{-1}\mathbf{A}_0 & -\mathbf{A}_2^{-1}\mathbf{A}_1 \end{bmatrix} \mathbf{T}(t) - \mathbf{T}^{-1}(t)\dot{\mathbf{T}}(t). \tag{3.55}$$

Both state companion matrices share the same eigenvalues because the algebraic multiplicity is preserved during decoupling. However, the corresponding geometric multiplicities might differ based on the differences between $\mathbf{J}_q$ and $\mathbf{J}_p$. In the event that $\mathbf{J}_p = \mathbf{J}_q$, the decoupling transformation is strictly isospectral and the real transformation matrix $\mathbf{T}$ does not depend on time t, hence these matrices are related by a similarity transformation and phase synchronization is structure-preserving. Time variation arises when not requiring that geometric multiplicity be preserved during decoupling [31, 32].

To see how this transformation is a generalization of previous results, suppose system (3.46) possesses distinct eigenvalues. Using primary–secondary pairing scheme,

$$\mathbf{J}_q = \begin{bmatrix} \boldsymbol{\Lambda}_1 & \mathbf{0} \\ \mathbf{0} & \boldsymbol{\Lambda}_2 \end{bmatrix}, \quad \mathbf{V}_q = \begin{bmatrix} \mathbf{V}_1 & \mathbf{V}_2 \end{bmatrix}. \tag{3.56}$$

The decoupled system (3.49) implies that

$$\mathbf{J}_p = \mathbf{J}_q, \quad \mathbf{V}_p = \begin{bmatrix} \mathbf{I} & \mathbf{I} \end{bmatrix}, \tag{3.57}$$

where $\mathbf{I}$ is the n-dimensional identity matrix. As a result, the transformation in (3.52) reduces to the transformation in Eq. (3.18) taking $\mathbf{f}(t) = \mathbf{0}$. Further, when the system is classically damped, this transformation reduces to the modal transformation $\mathbf{q} = \mathbf{U}\mathbf{p}$.

3.3.1.1 Example 3.4

Consider a non-classically damped system

$$\ddot{\mathbf{q}} + \begin{bmatrix} 2 & -1 \\ -1 & 2 \end{bmatrix} \dot{\mathbf{q}} + \begin{bmatrix} 2 & -1 \\ -1 & 5 \end{bmatrix} \mathbf{q} = \mathbf{0} \tag{3.58}$$

with $\mathbf{q}(0) = [1 \quad 1]^T$ and $\dot{\mathbf{q}}(0) = \mathbf{0}$. This system possesses a defective complex eigenvalue $\lambda_1 = -1 + i\sqrt{2}$ with algebraic multiplicity of 2 and geometric multiplicity of 1. This means an eigenvector and generalized eigenvector can be determined from Eq. (1.24):

$$\mathbf{v}_1 = \begin{bmatrix} -i\sqrt{2} \\ 1 \end{bmatrix}, \quad \mathbf{v}_1^2 = \begin{bmatrix} 3 \\ 0 \end{bmatrix}. \tag{3.59}$$

Its complex conjugate $\lambda_2 = -1 - i\sqrt{2}$ is associated with $\mathbf{v}_2 = \overline{\mathbf{v}}_1$ and $\mathbf{v}_2^2 = \mathbf{v}_1^2$. These can be assembled in

$$\mathbf{J}_q = \begin{bmatrix} -1+i\sqrt{2} & 1 & 0 & 0 \\ 0 & -1+i\sqrt{2} & 0 & 0 \\ 0 & 0 & -1-i\sqrt{2} & 1 \\ 0 & 0 & 0 & -1-i\sqrt{2} \end{bmatrix}, \tag{3.60}$$

$$\mathbf{V}_q = \begin{bmatrix} -i\sqrt{2} & 3 & i\sqrt{2} & 3 \\ 1 & 0 & 1 & 0 \end{bmatrix}. \tag{3.61}$$

Since complex conjugate eigensolutions must be paired together,

$$\mathbf{\Lambda}_1 = (-1 + i\sqrt{2})\mathbf{I}, \ \mathbf{\Lambda}_2 = (-1 - i\sqrt{2})\mathbf{I}. \tag{3.62}$$

Decoupling is achieved by setting

$$\mathbf{J}_p = \begin{bmatrix} \mathbf{\Lambda}_1 & \mathbf{0} \\ \mathbf{0} & \mathbf{\Lambda}_2 \end{bmatrix}, \quad \mathbf{V}_p = \begin{bmatrix} 1 & 0 & 1 & 0 \\ 0 & 1 & 0 & 1 \end{bmatrix} \tag{3.63}$$

so the real decoupled system is composed of two equal underdamped degrees of freedom:

$$\ddot{\mathbf{p}} + 2\mathbf{I}\dot{\mathbf{p}} + 3\mathbf{I}\mathbf{p} = \mathbf{0}. \tag{3.64}$$

Note that the coupled and decoupled systems are not strictly isospectral, and the decoupling transformation (3.52) is time-dependent:

$$\mathbf{T}(t) = \begin{bmatrix} -1 & 3-t & -1 & -t \\ 1 & t & 0 & 0 \\ 3 & 3t-1 & 1 & t+2 \\ 0 & 1 & 1 & t \end{bmatrix} \tag{3.65}$$

and the initial conditions of the decoupled system are

$$\begin{bmatrix} \mathbf{p}(0) \\ \dot{\mathbf{p}}(0) \end{bmatrix} = \mathbf{T}^{-1}(0) \begin{bmatrix} \mathbf{q}(0) \\ \dot{\mathbf{q}}(0) \end{bmatrix} = \begin{bmatrix} 1 \\ 1/2 \\ -1/2 \\ -1 \end{bmatrix}. \tag{3.66}$$

The configuration-space transformation is

$$\mathbf{q}(t) = \begin{bmatrix} -1 & 3-t \\ 1 & t \end{bmatrix} \mathbf{p}(t) + \begin{bmatrix} -1 & -t \\ 0 & 0 \end{bmatrix} \dot{\mathbf{p}}(t). \tag{3.67}$$

The response is shown in Fig. 3.5. This example shows how decoupling defective systems by the method of phase synchronization is straightforward, but laborious. Intuitively speaking, the phase synchronization decoupling transformation (3.52) is analogous to introducing different phase shifts among different Jordan blocks.

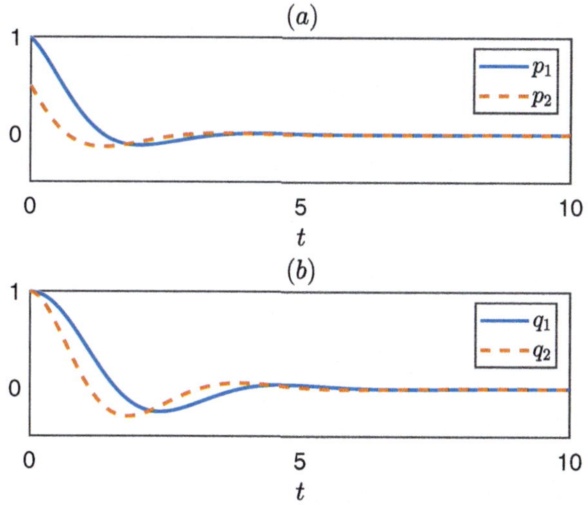

Fig. 3.5 (a) Response of the decoupled system and (b) response of coupled system of Example 3.4. These two systems are not strictly isospectral.

3.3.2 Phase Synchronization of Forced Vibrations

Consider an inhomogeneous system of the form

$$\mathbf{A}_2\ddot{\mathbf{q}} + \mathbf{A}_1\dot{\mathbf{q}} + \mathbf{A}_0\mathbf{q} = \mathbf{f}(t). \tag{3.68}$$

The objective is to decouple Eq. (3.68) into

$$\ddot{\mathbf{p}} + \mathbf{D}\dot{\mathbf{p}} + \boldsymbol{\Omega}\mathbf{p} = \mathbf{g}(t), \tag{3.69}$$

where $\mathbf{D} = -\text{diag}[\lambda_1 + \lambda_{n+1}, \ldots, \lambda_n + \lambda_{2n}]$ and $\boldsymbol{\Omega} = \text{diag}[\lambda_1\lambda_{n+1}, \ldots, \lambda_n\lambda_{2n}]$ with appropriate pairing of the eigensolutions λ_j, λ_{n+j} $(j = 1, 2, \ldots, n)$. Consider the state-space transformation [82]

$$\begin{bmatrix} \mathbf{q} \\ \dot{\mathbf{q}} \end{bmatrix} = \mathbf{T}(t) \begin{bmatrix} \mathbf{p}_1 \\ \mathbf{p}_2 \end{bmatrix}, \tag{3.70}$$

where $\mathbf{T}(t)$ is given as in Eq. (3.52). After casting (3.68) in state space as

$$\begin{bmatrix} \dot{\mathbf{q}} \\ \ddot{\mathbf{q}} \end{bmatrix} = \begin{bmatrix} \mathbf{0} & \mathbf{I} \\ -\mathbf{A}_2^{-1}\mathbf{A}_0 & -\mathbf{A}_2^{-1}\mathbf{A}_1 \end{bmatrix} \begin{bmatrix} \mathbf{q} \\ \dot{\mathbf{q}} \end{bmatrix} + \begin{bmatrix} \mathbf{0} \\ \mathbf{A}_2^{-1}\mathbf{f}(t) \end{bmatrix}, \tag{3.71}$$

substitute Eq. (3.70) into (3.71), pre-multiply the result by $\mathbf{T}^{-1}(t)$ and use relationship (3.55) to obtain

$$\begin{bmatrix} \dot{\mathbf{p}}_1 \\ \dot{\mathbf{p}}_2 \end{bmatrix} = \begin{bmatrix} \mathbf{0} & \mathbf{I} \\ -\boldsymbol{\Omega} & -\mathbf{D} \end{bmatrix} \begin{bmatrix} \mathbf{p}_1 \\ \mathbf{p}_2 \end{bmatrix} + \begin{bmatrix} \mathbf{g}_1(t) \\ \mathbf{g}_2(t) \end{bmatrix}, \tag{3.72}$$

where

$$\begin{bmatrix} \mathbf{g}_1(t) \\ \mathbf{g}_2(t) \end{bmatrix} = \mathbf{T}^{-1}(t) \begin{bmatrix} \mathbf{0} \\ \mathbf{A}_2^{-1}\mathbf{f}(t) \end{bmatrix} = \begin{bmatrix} \mathbf{G}_1(t)\mathbf{f}(t) \\ \mathbf{G}_2(t)\mathbf{f}(t) \end{bmatrix}. \tag{3.73}$$

Extracting the upper-half of Eq. (3.72), eliminating coordinate $\mathbf{p}_2$, and comparing the result with Eq. (3.69) reveals that $\mathbf{p}_1(t) = \mathbf{p}(t)$ and the excitation $\mathbf{g}(t)$ is

$$\mathbf{g}(t) = \left(\mathbf{D} + \mathbf{I}\frac{d}{dt}\right)\mathbf{G}_1(t)\mathbf{f}(t) + \mathbf{G}_2(t)\mathbf{f}(t). \tag{3.74}$$

Consequently, from Eq. (3.70), the decoupling transformation in state space is

$$\begin{bmatrix} \mathbf{p} \\ \dot{\mathbf{p}} \end{bmatrix} = \mathbf{T}^{-1}(t) \begin{bmatrix} \mathbf{q} \\ \dot{\mathbf{q}} \end{bmatrix} + \begin{bmatrix} \mathbf{0} \\ \mathbf{G}_1(t)\mathbf{f}(t) \end{bmatrix}. \tag{3.75}$$

When $t = 0$, the above equation generates the initial values $\mathbf{p}(0)$ and $\dot{\mathbf{p}}(0)$. The corresponding configuration-space decoupling transformation is given by

$$\mathbf{q}(t) = \left[\mathbf{T}_1(t) + \mathbf{T}_2(t)\frac{d}{dt}\right]\mathbf{p}(t) - \mathbf{T}_2(t)\mathbf{G}_1(t)\mathbf{f}(t). \tag{3.76}$$

3.3.2.1 Example 3.5

A two-degree-of-freedom system is governed by

$$\ddot{\mathbf{q}} + \begin{bmatrix} 2 & -1 \\ -1 & 2 \end{bmatrix}\dot{\mathbf{q}} + \begin{bmatrix} 5 & -1 \\ -1 & 10 \end{bmatrix}\mathbf{q} = \begin{bmatrix} 1 \\ -2 \end{bmatrix}\cos t \tag{3.77}$$

with $\mathbf{q}(0) = \mathbf{0}$ and $\dot{\mathbf{q}}(0) = \mathbf{0}$. This system possesses a defective complex eigenvalue $\lambda_1 = -1 + i\sqrt{6}$ with eigenvector and generalized eigenvector

$$\mathbf{v}_1^1 = \begin{bmatrix} -i\sqrt{6}/2 \\ 1 \end{bmatrix}, \quad \mathbf{v}_1^2 = \begin{bmatrix} 5/2 \\ 1 \end{bmatrix}. \tag{3.78}$$

Assemble the arrays

$$\mathbf{J}_q = \begin{bmatrix} -1+i\sqrt{6} & 1 & 0 & 0 \\ 0 & -1+i\sqrt{6} & 0 & 0 \\ 0 & 0 & -1-i\sqrt{6} & 1 \\ 0 & 0 & 0 & -1-i\sqrt{6} \end{bmatrix}, \tag{3.79}$$

$$\mathbf{V}_q = \begin{bmatrix} -i\sqrt{6}/2 & 5/2 & i\sqrt{6}/2 & 5/2 \\ 1 & 0 & 1 & 0 \end{bmatrix}, \tag{3.80}$$

$$\mathbf{\Lambda}_1 = \bar{\mathbf{\Lambda}}_2 = (-1 + i\sqrt{6})\mathbf{I}, \tag{3.81}$$

$$\mathbf{J}_p = \begin{bmatrix} \mathbf{\Lambda}_1 & \mathbf{0} \\ \mathbf{0} & \mathbf{\Lambda}_2 \end{bmatrix}, \tag{3.82}$$

$$\mathbf{V}_p = \begin{bmatrix} 1 & 0 & 1 & 0 \\ 0 & 1 & 0 & 1 \end{bmatrix}. \tag{3.83}$$

Next, construct the transformation matrix

$$\mathbf{T}(t) = \begin{bmatrix} \mathbf{V}_q \\ \mathbf{V}_q\mathbf{J}_q \end{bmatrix} e^{\mathbf{J}_q t} e^{-\mathbf{J}_p t} \begin{bmatrix} \mathbf{V}_p \\ \mathbf{V}_p\mathbf{J}_p \end{bmatrix}^{-1}$$
$$= \begin{bmatrix} -1/2 & 5/2 - t/2 & -1/2 & -t/2 \\ 1 & t & 0 & 0 \\ 7/2 & 7t/2 - 1/2 & 1/2 & t/2 + 2 \\ 0 & 1 & 1 & t \end{bmatrix} \tag{3.84}$$

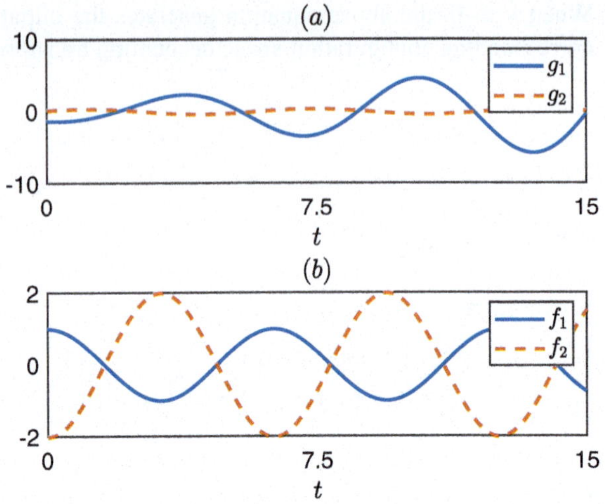

Fig. 3.6 (**a**) Driving force $\mathbf{g}(t)$ of decoupled system and (**b**) driving force $\mathbf{f}(t)$ of coupled system of Example 3.5. Since the decoupling process is not strictly isospectral, the time factors that would appear in $e^{\mathbf{J}_q t}$ are arrested in $\mathbf{g}(t)$ and the time-dependent transformation matrix $\mathbf{T}(t)$

and the matrices given by Eq. (3.73):

$$\mathbf{G}_1(t) = \begin{bmatrix} 0 & -t/6 \\ 0 & 1/6 \end{bmatrix}, \tag{3.85}$$

$$\mathbf{G}_2(t) = \begin{bmatrix} -t/2 & t/6 + 5/6 \\ 1/2 & -1/6 \end{bmatrix}. \tag{3.86}$$

This means the forcing vector $\mathbf{g}(t)$ of the decoupled system given by Eq. (3.74) is

$$\mathbf{g}(t) = \begin{bmatrix} -4\cos(t)/3 - t\cos(t)/6 - t\sin(t)/3 \\ \cos(t)/6 + \sin(t)/3 \end{bmatrix}. \tag{3.87}$$

The driving force is illustrated in Fig. 3.6.

The decoupled system is given by

$$\ddot{\mathbf{p}} + 2\mathbf{I}\dot{\mathbf{p}} + 7\mathbf{I}\mathbf{p} = \mathbf{g}(t) \tag{3.88}$$

with

$$\begin{bmatrix} \mathbf{p}(0) \\ \dot{\mathbf{p}}(0) \end{bmatrix} = \begin{bmatrix} 0 \\ 0 \\ 0 \\ -1/3 \end{bmatrix}. \tag{3.89}$$

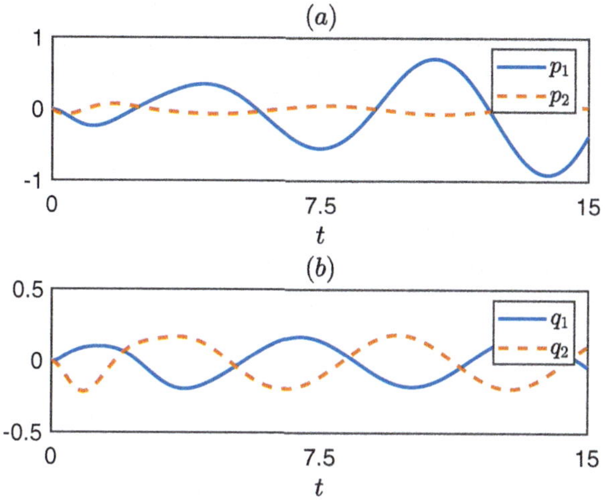

Fig. 3.7 (**a**) Response of the decoupled system and (**b**) response of coupled system of Example 3.5

The configuration-space transformation is

$$\mathbf{q}(t) = \begin{bmatrix} -1/2 & (5-t) \\ 1 & t \end{bmatrix} \mathbf{p}(t) + \begin{bmatrix} -1/2 & -t/2 \\ 0 & 0 \end{bmatrix} \dot{\mathbf{p}}(t) \tag{3.90}$$

because $\mathbf{T}_2\mathbf{G}_1 = \mathbf{0}$ .The response is shown in Fig. 3.7.

3.3.3 Decoupling by Isospectral Transformations

When a linear system possesses distinct eigenvalues, it is always decoupled by a strictly isospectral transformation. When a system admits repeated or defective eigenvalues, analysis requires attention. When can a linear system be decoupled by strictly isospectral transformation? In decoupling, the algebraic multiplicities of the eigenvalues are always preserved. Certain systems can be decoupled by strictly isospectral transformations, when one can choose $\mathbf{J}_p = \mathbf{J}_q$ and the decoupling transformation is time-invariant, while partial multiplicities of the eigenvalues may not be preserved for some other systems if decoupling is the ultimate goal.

Lancaster and Zaballa [37], and later Zuñiga Anaya [83], determined necessary and sufficient conditions under which strictly isospectral decoupling is possible. Recently, Kawano et al. [32] expressed these conditions in a more convenient form.

Theorem 3.1 *Any second-order linear system with invertible leading coefficient and whose associated spectrum satisfies*

1. *all complex eigenvalues are semi-simple and occur in conjugate pairs;*
2. *all Jordan blocks associated with a defective eigenvalue are no larger than* 2×2*;*
3. *excluding the complex eigenvalues and the* 2×2 *Jordan blocks, all remaining real eigenvalues, which have unit partial multiplicities, form pairs of distinct eigenvalues;*

can be converted into a real diagonal form by a linear time-invariant transformation that preserves the eigenvalues and their partial multiplicities.

Theorem 3.1 can be interpreted as follows. Consider the j-th component equation of the decoupled system:

$$\ddot{p}_j - (\lambda_j + \lambda_{n+j})\dot{p}_j + \lambda_j\lambda_{n+j}p_j = g_j(t). \tag{3.91}$$

This single-degree-of-freedom system has real coefficients and the roots of its characteristic polynomial λ_j and λ_{n+j} are either complex conjugates, distinct and real, or defective and real (forming a Jordan block of size 2×2). The limitations on the spectrum of each of these component equations make it not possible for the entirety of the spectrum of the coupled system to be represented by the congregation of the single-degree-of-freedom systems that compose the decoupled system. Therefore a two-degree-of-freedom system that has a defective complex eigenvalue cannot have an associated decoupled system with this eigenstructure. The best one can do in this case is to decouple with the eigenvalues preserved, but not their partial multiplicities (see Example 3.4 and Example 3.5 where the decoupled system was formed by two equal damped oscillators while the coupled system has a defective complex conjugate pair).

If conditions of Theorem 3.1 are not met, one can still decouple by only preserving the algebraic multiplicities of the eigenvalues. A flowchart for decoupling systems with repeated or defective eigenvalues is shown in Fig. 3.8

3.3.3.1 Example 3.6

Consider the system

$$\ddot{\mathbf{q}} + \begin{bmatrix} 2 & -1 \\ -1 & 3 \end{bmatrix}\dot{\mathbf{q}} + \begin{bmatrix} 1 & -1 \\ -1 & 3 \end{bmatrix}\mathbf{q} = \mathbf{0}. \tag{3.92}$$

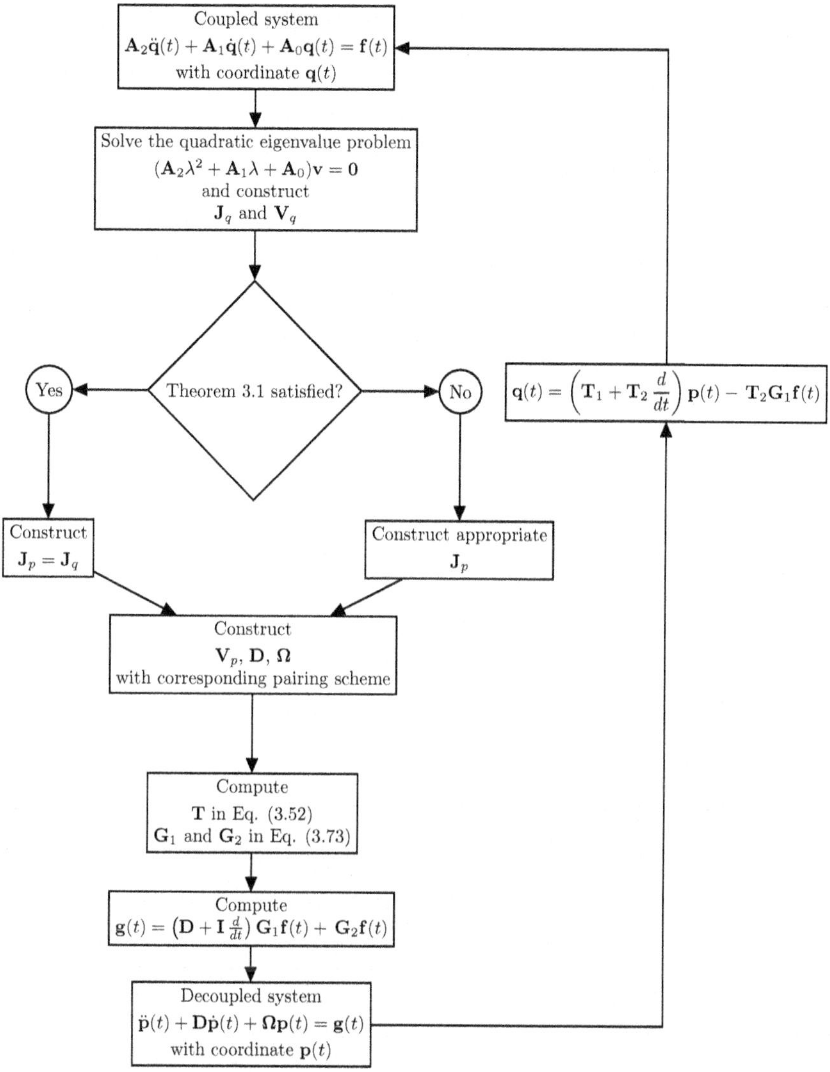

Fig. 3.8 Flowchart for decoupling by phase synchronization a linear system possessing repeated or defective eigenvalues. If Theorem 3.1 is satisfied, strictly isospectral decoupling is permissible and the transformation matrices $\mathbf{T}$, $\mathbf{G}_1$ and $\mathbf{G}_2$ are time-independent. Otherwise, these matrices may depend on time. Decoupling systems with repeated or defective eigenvalues is not as straightforward as dealing with systems with distinct eigenvalues

This system has one simple real eigenvalue and a defective real eigenvalue:

$$\mathbf{J}_q = \begin{bmatrix} -2 & 0 & 0 & 0 \\ 0 & -1 & 1 & 0 \\ 0 & 0 & -1 & 1 \\ 0 & 0 & 0 & -1 \end{bmatrix}, \tag{3.93}$$

$$\mathbf{V}_q = \begin{bmatrix} -1 & 1 & 1 & 0 \\ 1 & 0 & 1 & 1 \end{bmatrix}. \tag{3.94}$$

Since the associated Jordan block has dimension 3, conditions in Theorem 3.1 are not met and strictly isospectral decoupling is not possible. However, decoupling is still permissible if preservation of partial multiplicities is not required. A decoupled system can be obtained by pairing the distinct eigenvalue with any of the repeated eigenvalues. Picking

$$\mathbf{J}_p = \begin{bmatrix} -2 & 0 & 0 & 0 \\ 0 & -1 & 0 & 0 \\ 0 & 0 & -1 & 1 \\ 0 & 0 & 0 & -1 \end{bmatrix} \tag{3.95}$$

and

$$\mathbf{V}_p = \begin{bmatrix} 1 & 1 & 0 & 0 \\ 0 & 0 & 1 & 0 \end{bmatrix}, \tag{3.96}$$

the decoupled system is

$$\ddot{\mathbf{p}} + \begin{bmatrix} 3 & 0 \\ 0 & 2 \end{bmatrix} \dot{\mathbf{p}} + \begin{bmatrix} 2 & 0 \\ 0 & 1 \end{bmatrix} \mathbf{p} = \mathbf{0} \tag{3.97}$$

and the decoupling transformation is time-varying:

$$\mathbf{q}(t) = \begin{bmatrix} 3 & -t^2/2 + t + 2 \\ -1 & 1 \end{bmatrix} \mathbf{p}(t) + \begin{bmatrix} 2 & 1 - t^2/2 \\ -1 & 0 \end{bmatrix} \dot{\mathbf{p}}(t). \tag{3.98}$$

The response is shown in Fig. 3.9.

3.3.3.2 Example 3.7

Consider the non-symmetric system

$$\ddot{\mathbf{q}} + \begin{bmatrix} 5 & 1 \\ -1 & 2 \end{bmatrix} \dot{\mathbf{q}} + \begin{bmatrix} 4 & 1 \\ -1 & 1 \end{bmatrix} \mathbf{q} = \mathbf{0} \tag{3.99}$$

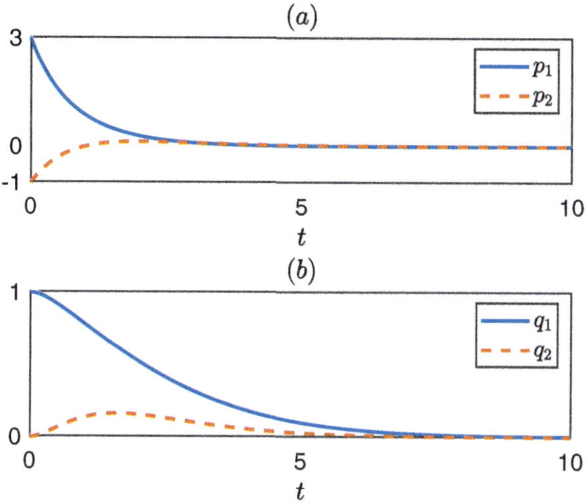

Fig. 3.9 (**a**) Response of the decoupled system and (**b**) response of coupled system of Example 3.6. The initial conditions are $\mathbf{q}(0) = [1 \quad 0]^T$ and $\dot{\mathbf{q}} = \mathbf{0}$

with $\mathbf{q}(0) = [1 \quad -1]^T$ and $\dot{\mathbf{q}}(0) = [-2 \quad -1]^T$. Solution of the quadratic eigenvalue problem (3.2) reveals this non-defective system possesses two simple eigenvalues and one eigenvalue repeated two times, but with linearly independent eigenvectors:

$$\mathbf{J}_q = \begin{bmatrix} -(5+\sqrt{5})/2 & 0 & 0 & 0 \\ 0 & -(5-\sqrt{5})/2 & 0 & 0 \\ 0 & 0 & -1 & 0 \\ 0 & 0 & 0 & -1 \end{bmatrix}, \tag{3.100}$$

$$\mathbf{V}_q = \begin{bmatrix} -(3+\sqrt{5})/2 & -(3-\sqrt{5})/2 & 1 & 0 \\ 1 & 1 & 0 & 1 \end{bmatrix}. \tag{3.101}$$

This system satisfies Theorem 3.1, and therefore can be decoupled by a strictly isospectral time-invariant transformation. To that end, the decoupled system is constructed by arranging the eigensolutions by primary–secondary scheme such that

$$\mathbf{\Lambda}_1 = \begin{bmatrix} -(5+\sqrt{5})/2 & 0 \\ 0 & -(5-\sqrt{5})/2 \end{bmatrix}, \ \mathbf{\Lambda}_2 = \begin{bmatrix} -1 & 0 \\ 0 & -1 \end{bmatrix}. \tag{3.102}$$

Decoupling is achieved by setting

$$\mathbf{J}_p = \mathbf{J}_q, \ \mathbf{V}_p = \begin{bmatrix} 1 & 0 & 1 & 0 \\ 0 & 1 & 0 & 1 \end{bmatrix}. \tag{3.103}$$

The decoupled system is then

$$\ddot{\mathbf{p}} + \begin{bmatrix} (7+\sqrt{5})/2 & 0 \\ 0 & (7-\sqrt{5})/2 \end{bmatrix} \dot{\mathbf{p}} + \begin{bmatrix} (5+\sqrt{5})/2 & 0 \\ 0 & (5-\sqrt{5})/2 \end{bmatrix} \mathbf{p} = \mathbf{0} \tag{3.104}$$

and the decoupling transformation (3.52) is

$$\mathbf{T} = \begin{bmatrix} (7-\sqrt{5})/2 & 1 & (5-\sqrt{5})/2 & 1 \\ (-3+\sqrt{5})/2 & 1 & (-3+\sqrt{5})/2 & 0 \\ -5 & (-5+\sqrt{5})/2 & -4 & (-5+\sqrt{5})/2 \\ (5-\sqrt{5})/2 & 0 & (5-\sqrt{5})/2 & 1 \end{bmatrix} \tag{3.105}$$

and

$$\mathbf{T}^{-1} = \begin{bmatrix} \dfrac{(5+3\sqrt{5})}{10} & \dfrac{(-25+7\sqrt{5})}{10} & \dfrac{(-5+3\sqrt{5})}{10} & \dfrac{(-25+7\sqrt{5})}{10} \\ \dfrac{(5-3\sqrt{5})}{10} & \dfrac{(25-7\sqrt{5})}{10} & \dfrac{(5-3\sqrt{5})}{10} & \dfrac{(15-7\sqrt{5})}{10} \\ -\dfrac{(1+\sqrt{5})}{2} & 3-\sqrt{5} & \dfrac{(1-\sqrt{5})}{2} & 3-\sqrt{5} \\ \dfrac{(-5+\sqrt{5})}{2} & -2+\sqrt{5} & \dfrac{(-1+\sqrt{5})}{2} & -1+\sqrt{5} \end{bmatrix}. \tag{3.106}$$

Therefore, the initial conditions of the decoupled system are

$$\begin{bmatrix} \mathbf{p}(0) \\ \dot{\mathbf{p}}(0) \end{bmatrix} = \mathbf{T}^{-1} \begin{bmatrix} \mathbf{q}(0) \\ \dot{\mathbf{q}}(0) \end{bmatrix} = \begin{bmatrix} (65-17\sqrt{5})/10 \\ (-45+17\sqrt{5})/10 \\ (-15+5\sqrt{5})/2 \\ (7-5\sqrt{5})/2 \end{bmatrix}. \tag{3.107}$$

The coordinates are related in configuration-space by the upper-half of Eq. (3.52):

$$\mathbf{q}(t) = \begin{bmatrix} (7+\sqrt{5})/2 & 1 \\ (-3+\sqrt{5})/2 & 1 \end{bmatrix} \mathbf{p}(t) + \begin{bmatrix} (5-\sqrt{5})/2 & 1 \\ (-3+\sqrt{5})/2 & 0 \end{bmatrix} \dot{\mathbf{p}}(t). \tag{3.108}$$

The response is shown in Fig. 3.10. It can be checked that

$$\begin{aligned} &\mathbf{T}^{-1} \begin{bmatrix} \mathbf{0} & \mathbf{I} \\ -\mathbf{A}_2^{-1}\mathbf{A}_0 & -\mathbf{A}_2^{-1}\mathbf{A}_1 \end{bmatrix} \mathbf{T} \\ &= \begin{bmatrix} 0 & 0 & 1 & 0 \\ 0 & 0 & 0 & 1 \\ -(5+\sqrt{5})/2 & 0 & -(7+\sqrt{5})/2 & 0 \\ 0 & (-5+\sqrt{5})/2 & 0 & (-7+\sqrt{5})/2 \end{bmatrix} \\ &= \begin{bmatrix} \mathbf{0} & \mathbf{I} \\ -\boldsymbol{\Omega} & -\mathbf{D} \end{bmatrix}. \end{aligned} \tag{3.109}$$

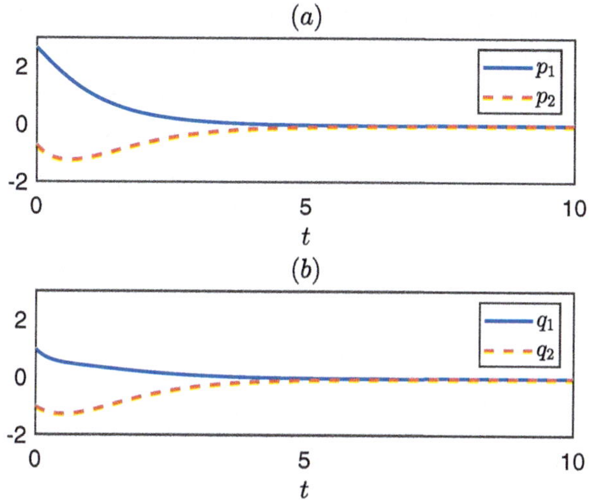

Fig. 3.10 (**a**) Response of the decoupled system and (**b**) response of coupled system of Example 3.7. Because $\mathbf{J}_p = \mathbf{J}_q$, the coupled and decoupled systems are strictly isospectral and the decoupling transformation in Eq. (3.52) is time-independent

Therefore the state transformation is structure-preserving.

The above decoupling procedure is strictly isospectral. The eigenvalues and partial multiplicities are preserved. Alternatively, this system can be decoupled by preserving only the eigenvalues, but not their geometric multiplicities if a different pairing scheme is used. To that end, the following indexing scheme is used:

$$\mathbf{\Lambda}_1 = \begin{bmatrix} -(5+\sqrt{5})/2 & 0 \\ 0 & -1 \end{bmatrix}, \quad \mathbf{\Lambda}_2 = \begin{bmatrix} -(5-\sqrt{5})/2 & 0 \\ 0 & -1 \end{bmatrix}. \tag{3.110}$$

The decoupled system is then composed of an overdamped and a critically damped degrees of freedom, the latter being defective:

$$\ddot{\mathbf{p}} + \begin{bmatrix} 5 & 0 \\ 0 & 2 \end{bmatrix} \dot{\mathbf{p}} + \begin{bmatrix} 5 & 0 \\ 0 & 1 \end{bmatrix} \mathbf{p} = \mathbf{0}. \tag{3.111}$$

In this case, decoupling is accomplished by

$$\mathbf{J}_p = \begin{bmatrix} -(5+\sqrt{5})/2 & 0 & 0 & 0 \\ 0 & -(5-\sqrt{5})/2 & 0 & 0 \\ 0 & 0 & -1 & 1 \\ 0 & 0 & 0 & -1 \end{bmatrix},$$

$$\mathbf{V}_p = \begin{bmatrix} 1 & 1 & 0 & 0 \\ 0 & 0 & 1 & 0 \end{bmatrix} \tag{3.112}$$

and the decoupling transformation (3.52) becomes time-dependent

$$\mathbf{T}(t) = \begin{bmatrix} 1 & 1-t & 1 & -t \\ 1 & 1 & 0 & 1 \\ -5 & t-1 & -4 & t \\ 0 & -1 & 1 & -1 \end{bmatrix},$$
$$\mathbf{T}^{-1}(t) = \begin{bmatrix} -1 & -3 & -1 & -3 \\ 1+t & -1+4t & t & -1+3t \\ 1 & 4 & 1 & 4 \\ -t & 5-4t & 1-t & 4-3t \end{bmatrix}, \tag{3.113}$$

because $\mathbf{J}_p \neq \mathbf{J}_q$. Therefore, the initial conditions of the decoupled system are

$$\begin{bmatrix} \mathbf{p}(0) \\ \dot{\mathbf{p}}(0) \end{bmatrix} = \mathbf{T}^{-1}(0) \begin{bmatrix} \mathbf{q}(0) \\ \dot{\mathbf{q}}(0) \end{bmatrix} = \begin{bmatrix} 7 \\ 3 \\ -9 \\ -11 \end{bmatrix}. \tag{3.114}$$

Finally,

$$\mathbf{q}(t) = \begin{bmatrix} 1 & 1-t \\ 1 & 1 \end{bmatrix} \mathbf{p}(t) + \begin{bmatrix} 1 & -t \\ 0 & 1 \end{bmatrix} \dot{\mathbf{p}}(t). \tag{3.115}$$

Even though strictly isospectral decoupling is only permissible when Theorem 3.1 is satisfied, decoupling by preserving only the algebraic multiplicities of the eigenvalues is always possible. In this case, the decoupling transformation satisfies

$$\begin{aligned} &\mathbf{T}^{-1} \begin{bmatrix} \mathbf{0} & \mathbf{I} \\ -\mathbf{A}_2^{-1}\mathbf{A}_0 & -\mathbf{A}_2^{-1}\mathbf{A}_1 \end{bmatrix} \mathbf{T} - \mathbf{T}^{-1}(t)\dot{\mathbf{T}}(t) \\ &= \begin{bmatrix} 0 & 0 & 1 & 0 \\ 0 & 0 & 0 & 1 \\ -5 & 0 & -5 & 0 \\ 0 & -1 & 0 & -2 \end{bmatrix} \\ &= \begin{bmatrix} \mathbf{0} & \mathbf{I} \\ -\boldsymbol{\Omega} & -\mathbf{D} \end{bmatrix}. \end{aligned} \tag{3.116}$$

3.4 Conclusions

Classical decoupling techniques focus on simultaneous diagonalization of the three system coefficients by time-independent transformations, while phase synchronization exploits time as a decoupling parameter. Several observations about the decoupling of linear systems by phase synchronization are summarized in the following remarks.

1. Significance of quadratic eigenvalue problem: in free or forced vibration, all parameters required for the decoupling of a linear system are obtained through the solution of a quadratic eigenvalue problem; generalized eigenvectors are used when eigenvalues are defective, and an admissible eigenvalue pairing scheme is needed.
2. Effect of normalization of eigenvectors: coefficient matrices $\mathbf{D}$ and $\boldsymbol{\Omega}$ of the decoupled system are independent of the normalization of eigenvectors because they are constructed from the eigenvalues. In contrast, the excitation $\mathbf{g}(t)$, the initial values $\mathbf{p}(0)$, $\dot{\mathbf{p}}(0)$, and the decoupling transformations in both configuration and state spaces are dependent on the eigenvectors and therefore on the normalization used.
3. Systems with distinct eigenvalues comprise the bulk of classical engineering applications. For these systems, the overall decoupling transformation is strictly isospectral, real, invertible, non-linear, and time-dependent in configuration space. A decoupling algorithm is provided in Fig. 3.2. It has been implemented in MATLAB to accompany this book and its use is illustrated in Appendix B.
4. Any system with repeated or defective eigenvalues can be decoupled if one does not insist on preserving geometric multiplicities, resulting in a possibly time-dependent decoupling transformation. Under conditions stated in Theorem 3.1, a dynamical system can be decoupled through a strictly isospectral transformation and the decoupling transformation is time-invariant.
5. Phase synchronization is a direct extension of modal analysis.

Chapter 4
Selected Applications

And we know that as long as men are free to ask what they must, free to say what they think, free to think what they will, freedom can never be lost, and science can never regress

—Julius Robert Oppenheimer

4.1 Characterization of Free Motion of Damped Systems

In many engineering applications, it is important to determine the effect of damping on the system response. For example, in engineering design applications one needs to know how oscillations can be suppressed by varying certain system parameters [84]. Characterization of the free motion of a damped single-degree-of-freedom system is well understood: the nature of damped motion can be determined by inspection of the viscous damping ratio. Consider a damped single-degree-of-freedom system:

$$m\ddot{q} + c\dot{q} + kq = 0, \tag{4.1}$$

where $m > 0$, $c > 0$, and $k > 0$. The damping ratio is a non-negative number defined by the coefficients of the system:

$$\zeta = \frac{c}{2\sqrt{mk}}. \tag{4.2}$$

The system is underdamped if $\zeta < 1$, it is overdamped if $\zeta > 1$, and it is critically damped if $\zeta = 1$. Oscillatory behavior is present whenever $\zeta < 1$.

The associated characteristic equation is

$$m\lambda^2 + c\lambda + k = 0, \tag{4.3}$$

whose roots λ_1 and λ_2 can be expressed as

$$\lambda_1, \lambda_2 = \frac{1}{2m}\left(-c \pm \sqrt{c^2 - 4mk}\right) = -\zeta\omega_n \pm \omega_n\sqrt{\zeta^2 - 1}, \tag{4.4}$$

R. Gonçalves Salsa Junior, F. Ma, *Advances in the Theory of System Decoupling*,
https://doi.org/10.1007/978-3-030-60846-0_4

where $\omega_n = \sqrt{k/m} > 0$ is the natural frequency. Underdamping ($\zeta < 1$) happens when λ_1 and λ_2 are complex conjugates. If the system is overdamped, $\zeta > 1$ and λ_1 and λ_2 are distinct negative real numbers. In case it is critically damped, $\zeta = 1$, in which case $\lambda_1 = \lambda_2 < 0$.

The damping ratio can also be expressed in terms of the roots of the characteristic equation as

$$\zeta = \frac{c/m}{2\sqrt{k/m}} = -\frac{\lambda_1 + \lambda_2}{2} \frac{1}{\sqrt{\lambda_1 \lambda_2}} = -\frac{\text{arithmetic mean of roots}}{\text{geometric mean of roots}}. \tag{4.5}$$

Here, the terms arithmetic and geometric means are used liberally because the roots may be complex. The inequality governing arithmetic and geometric means is valid only for positive numbers [85].

While a similar criterion for determining oscillatory behavior of the free motion of damped multi-degree-of-freedom systems is desired, the situation here is less clear. A sufficient condition for non-oscillatory behavior is Duffin's overdamping condition, represented by Eq. (2.5) [86]. When the overdamping condition is satisfied, all eigenvalues are negative and the free response is exponentially decaying [59]. However, the overdamping condition is only sufficient and not necessary for overdamping. Other conditions reported in [87–89] are rather difficult to verify and have not really found their ways into applications. Morzfeld et al. [84] introduced a viscous damping function that represents a direct extension of the classical damping ratio. The effect of viscous damping on the free motion is then determined by minimization and maximization of this viscous damping function. However, optimization of such function may be problematic because the iterations can get trapped around local extrema in applications. Other approaches to this problem [90–94] rely upon simultaneous diagonalization of the coefficient matrices by linear coordinate transformations: these techniques apply to classically damped systems only and are not applicable in general.

4.1.1 Characterization of Free Motion by Eigenvalues

Suppose an n-degree-of-freedom is decoupled by phase synchronization. The decoupled coordinate p_j ($j = 1, 2, \ldots, n$) satisfies a second-order differential equation of the form:

$$\ddot{p}_j - (\lambda_j + \lambda_{n+j})\dot{p}_j + (\lambda_j \lambda_{n+j}) p_j = 0, \tag{4.6}$$

where $\lambda_{n+j} = \bar{\lambda}_j$ whenever λ_j is complex. Recall that the eigenvalues of a damped system are complex conjugates with either negative real part or real and negative. As in the case of single-degree-of-freedom systems (4.5), denote the damping ratio of the jth scalar equation of the decoupled system by

$$\zeta_{pj} = -\frac{\lambda_j + \lambda_{n+j}}{2} \frac{1}{\sqrt{\lambda_j \lambda_{n+j}}}. \tag{4.7}$$

When λ_j is complex, $\lambda_{n+j} = \bar{\lambda}_j$ and $\zeta_{pj} < 1$. This means each pair of complex conjugate eigenvalues produces an underdamped degree of freedom in the decoupled system. When λ_j and λ_{n+j} are real eigenvalues, they produce an overdamped degree of freedom when $\lambda_j \neq \lambda_{n+j}$ or a critically damped degree of freedom when $\lambda_j = \lambda_{n+j}$. This leads to the conclusion in Theorem 4.1:

Theorem 4.1 *If there are $2n_c$ complex conjugate eigenvalues and $2n_r = 2(n - n_c)$ real eigenvalues, then n_c degrees of freedom are underdamped and n_r degrees of freedom are overdamped or critically damped.*

Because of coupling, the system response is oscillatory if at least one scalar equation in the decoupled system is underdamped. This is an exact and complete characterization, but knowledge of the eigenvalues is required.

4.1.1.1 Example 4.1

Consider the three-degree-of-freedom system:

$$\ddot{\mathbf{q}} + \begin{bmatrix} 12 & 10 & 0 \\ 10 & 12 & -1 \\ 0 & -1 & 2 \end{bmatrix} \dot{\mathbf{q}} + \begin{bmatrix} 12 & 11 & 6 \\ 11 & 12 & 5 \\ 6 & 5 & 30 \end{bmatrix} \mathbf{q} = \mathbf{0}. \tag{4.8}$$

Solution of the quadratic eigenvalue problem reveals that the spectrum consists of:

- a complex conjugate pair $\lambda_1 = -1.0330 + 5.4034i$, $\lambda_4 = \bar{\lambda}_1$ with eigenvectors:

$$\mathbf{v}_1 = \begin{bmatrix} 1 \\ -0.2441 - 0.3290i \\ 1.8187 - 6.714i \end{bmatrix}, \quad \mathbf{v}_4 = \bar{\mathbf{v}}_1. \tag{4.9}$$

- two simple real eigenvalues, $\lambda_2 = -20.9597$ and $\lambda_6 = -0.9743$ with eigenvectors:

$$\mathbf{v}_2 = \begin{bmatrix} 1 \\ 1.0038 \\ -0.0750 \end{bmatrix}, \quad \mathbf{v}_6 = \begin{bmatrix} 1 \\ -0.7817 \\ -0.0459 \end{bmatrix}. \tag{4.10}$$

- a defective real eigenvalue repeated two times: $\lambda_3 = -1$. This eigenvalue has unit geometric multiplicity and its eigenvector is

$$\mathbf{v}_3^1 = \begin{bmatrix} 1 \\ -1 \\ 0 \end{bmatrix}. \tag{4.11}$$

The associated generalized eigenvector $\mathbf{v}_3^2$, given by (1.24), is

$$\mathbf{v}_3^2 = \begin{bmatrix} 1 \\ -1.8571 \\ 0.1429 \end{bmatrix}. \tag{4.12}$$

Note that each eigenvector was normalized so that the first element is 1. Conditions of Theorem 3.1 are met, so the indexing of eigenvalues

$$\begin{aligned} \mathbf{\Lambda}_1 &= \begin{bmatrix} -1.0330+5.4034i & 0 & 0 \\ 0 & -20.9597 & 0 \\ 0 & 0 & -1 \end{bmatrix}, \\ \mathbf{\Lambda}_2 &= \begin{bmatrix} -1.0330-5.4034i & 0 & 0 \\ 0 & -0.9743 & 0 \\ 0 & 0 & -1 \end{bmatrix} \end{aligned} \tag{4.13}$$

culminate in the decoupled system:

$$\ddot{\mathbf{p}} + \begin{bmatrix} 2.0660 & 0 & 0 \\ 0 & 21.9340 & 0 \\ 0 & 0 & 2 \end{bmatrix} \dot{\mathbf{p}} + \begin{bmatrix} 30.2638 & 0 & 0 \\ 0 & 20.4210 & 0 \\ 0 & 0 & 1 \end{bmatrix} \mathbf{p} = \mathbf{0}, \tag{4.14}$$

where the complex eigenvalues produce an underdamped degree of freedom with $\zeta_{p1} = 0.1878$, while the distinct real eigenvalues originate an overdamped degree of freedom with $\zeta_{p2} = 2.4269$. The defective real eigenvalue forms a critically damped degree of freedom with $\zeta_{p3} = 1$. To compute the decoupling transformation, let

$$\mathbf{J}_q = \begin{bmatrix} \lambda_1 & 0 & 0 & 0 & 0 & 0 \\ 0 & \bar{\lambda}_1 & 0 & 0 & 0 & 0 \\ 0 & 0 & \lambda_2 & 0 & 0 & 0 \\ 0 & 0 & 0 & \lambda_6 & 0 & 0 \\ 0 & 0 & 0 & 0 & \lambda_3 & 1 \\ 0 & 0 & 0 & 0 & 0 & \lambda_3 \end{bmatrix}, \quad \mathbf{V}_q = \begin{bmatrix} \mathbf{v}_1 & \bar{\mathbf{v}}_1 & \mathbf{v}_2 & \mathbf{v}_6 & \mathbf{v}_3^1 & \mathbf{v}_3^2 \end{bmatrix}. \tag{4.15}$$

For isospectral decoupling, $\mathbf{J}_p = \mathbf{J}_q$ and define the matrix:

$$\mathbf{V}_p = \begin{bmatrix} 1 & 1 & 0 & 0 & 0 & 0 \\ 0 & 0 & 1 & 1 & 0 & 0 \\ 0 & 0 & 0 & 0 & 1 & 0 \end{bmatrix}. \tag{4.16}$$

The decoupling transformation is then given by

$$\begin{aligned} \mathbf{q}(t) &= \left(\mathbf{T}_1 + \mathbf{T}_2 \frac{d}{dt}\right) \mathbf{p}(t) \\ &= \begin{bmatrix} 1 & 1 & 2 \\ -0.3070 & -0.8687 & -2.8571 \\ 0.5357 & -0.0444 & 0.1429 \end{bmatrix} \begin{bmatrix} p_1(t) \\ p_2(t) \\ p_3(t) \end{bmatrix} \\ &+ \begin{bmatrix} 0 & 0 & 1 \\ -0.0609 & -0.0893 & -1.8571 \\ -1.2421 & 0.0015 & 0.1429 \end{bmatrix} \begin{bmatrix} \dot{p}_1(t) \\ \dot{p}_2(t) \\ \dot{p}_3(t) \end{bmatrix}. \end{aligned} \tag{4.17}$$

The responses of coupled and decoupled systems are shown in Fig. 4.1.

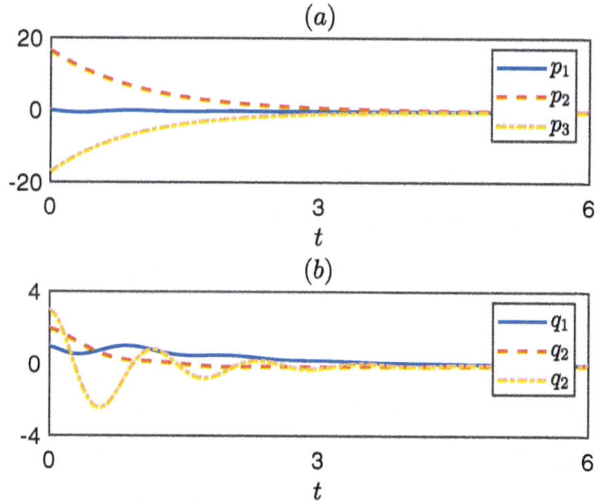

Fig. 4.1 (**a**) the response $\mathbf{p}(t)$ of the decoupled system and (**b**) the response $\mathbf{q}(t)$ of the coupled system of Example 4.1. Complex conjugate eigenvalues originate an underdamped degree of freedom with $\zeta_{p1} = 0.1878$, distinct real eigenvalues originate an overdamped degree of freedom with $\zeta_{p2} = 2.4269$, and repeated real eigenvalues originate a critically damped degree of freedom with $\zeta_{p3} = 1$. The initial conditions used for the simulation are $\mathbf{q}(0) = [1 \quad 2 \quad 3]^T$ and $\dot{\mathbf{q}}(0) = \mathbf{0}$

Note that the eigenvalues can also be paired in accordance with

$$\boldsymbol{\Lambda}_1 = \begin{bmatrix} -1.0330+5.4034i & 0 & 0 \\ 0 & -20.9597 & 0 \\ 0 & 0 & -1 \end{bmatrix}, \qquad (4.18)$$

$$\boldsymbol{\Lambda}_2 = \begin{bmatrix} -1.0330-5.4034i & 0 & 0 \\ 0 & -1 & 0 \\ 0 & 0 & -0.9743 \end{bmatrix}$$

and the decoupled system becomes

$$\ddot{\mathbf{p}} + \begin{bmatrix} 2.0660 & 0 & 0 \\ 0 & 21.9597 & 0 \\ 0 & 0 & 1.9743 \end{bmatrix} \dot{\mathbf{p}} + \begin{bmatrix} 30.2638 & 0 & 0 \\ 0 & 20.9597 & 0 \\ 0 & 0 & 0.9743 \end{bmatrix} \mathbf{p} = \mathbf{0}. \qquad (4.19)$$

The complex conjugate eigenvalues produce an underdamped degree of freedom, but the pairing of real eigenvalues produces two overdamped degrees of freedom with $\zeta_{p2} = 2.3983$ and $\zeta_{p3} = 1.0001$. To obtain the decoupling transformation, let

$$\mathbf{J}_p = \begin{bmatrix} \lambda_1 & 0 & 0 & 0 & 0 & 0 \\ 0 & \bar{\lambda}_1 & 0 & 0 & 0 & 0 \\ 0 & 0 & \lambda_2 & 0 & 0 & 0 \\ 0 & 0 & 0 & \lambda_6 & 0 & 0 \\ 0 & 0 & 0 & 0 & \lambda_3 & 0 \\ 0 & 0 & 0 & 0 & 0 & \lambda_3 \end{bmatrix}, \quad \mathbf{V}_p = \begin{bmatrix} 1 & 1 & 0 & 0 & 0 & 0 \\ 0 & 0 & 1 & 0 & 1 & 0 \\ 0 & 0 & 0 & 1 & 0 & 1 \end{bmatrix}. \qquad (4.20)$$

The decoupling transformation is then given by

$$\begin{aligned} \mathbf{q}(t) &= \left(\mathbf{T}_1 + \mathbf{T}_2 \frac{d}{dt}\right) \mathbf{p}(t) \\ &= \begin{bmatrix} 1 & 1 & 1-37.8524t \\ -0.3070 & -1.1004 & 37.8524t+39.9263 \\ 0.5357 & 0.0038 & -7.1892 \end{bmatrix} \begin{bmatrix} p_1(t) \\ p_2(t) \\ p_3(t) \end{bmatrix} \\ &+ \begin{bmatrix} 0 & 0 & -38.8524t \\ -0.0609 & -1.1004 & 38.8524t+41.7835 \\ -1.2421 & 0.0038 & -7.3320 \end{bmatrix} \begin{bmatrix} \dot{p}_1(t) \\ \dot{p}_2(t) \\ \dot{p}_3(t) \end{bmatrix}. \end{aligned} \qquad (4.21)$$

4.1.1.2 Example 4.2

Consider a damped system with

$$\mathbf{M} = \mathbf{I}, \quad \mathbf{C} = \begin{bmatrix} 19 & 9 \\ 9 & 19 \end{bmatrix}, \quad \mathbf{K} = \begin{bmatrix} 91 & 70 \\ 70 & 91 \end{bmatrix}. \tag{4.22}$$

The eigenvalues can be arranged into a primary–secondary scheme as

$$\mathbf{\Lambda}_1 = \begin{bmatrix} \lambda_1 & 0 \\ 0 & \lambda_2 \end{bmatrix} = \begin{bmatrix} -19.9161 & 0 \\ 0 & -8.0839 \end{bmatrix}, \tag{4.23}$$

$$\mathbf{\Lambda}_2 = \begin{bmatrix} \lambda_3 & 0 \\ 0 & \lambda_4 \end{bmatrix} = \begin{bmatrix} -7 & 0 \\ 0 & -3 \end{bmatrix}. \tag{4.24}$$

In this system, all eigenvalues are negative and distinct; therefore, all degrees of freedom are overdamped. For the input $[0 \quad 1]^T$, Duffin's condition (2.5) gives

$$\Delta([0 \quad 1]^T) = -3. \tag{4.25}$$

The overdamping condition is not satisfied, even though each degree of freedom is overdamped. Damping criteria involving functionals are not convenient to use. A criterion that utilizes a real number to determine if the system response is oscillatory is preferred.

4.1.2 Characterization of Free Motion by Damping Ratio

When an n-degree-of-freedom is decoupled by phase synchronization, the damping ratio of the jth degree of freedom is given by (4.7). Since different schemes can be used to pair real eigensolutions, parameters to characterize damping for the coupled system such as

$$\zeta_p = \zeta_{p1}\zeta_{p2}\ldots\zeta_{pn}, \tag{4.26}$$

$$\zeta_{\max} = \max_{1\le j\le n} \zeta_{pj}, \tag{4.27}$$

or

$$\zeta_{\min} = \min_{1\le j\le n} \zeta_{pj} \tag{4.28}$$

are not fixed and their values are dependent on the chosen pairing scheme. Moreover, the computation of these parameters or even each of ζ_{pj} $(j = 1, 2, \ldots, n)$ depends on the solution of the quadratic eigenvalue problem. A parameter that depends only on the given system's parameters, such as the coefficient matrices, is desired.

Define a system damping ratio by

$$\zeta = \frac{\operatorname{tr}\left(\mathbf{M}^{-1}\mathbf{C}\right)}{2n} \frac{1}{\sqrt[2n]{\det\left(\mathbf{M}^{-1}\mathbf{K}\right)}}, \tag{4.29}$$

where $\operatorname{tr}\left(\mathbf{M}^{-1}\mathbf{C}\right)$ is the trace of $\mathbf{M}^{-1}\mathbf{C}$. Since $\mathbf{M}$, $\mathbf{C}$, and $\mathbf{K}$ are positive-definite, the system damping ratio ζ is a non-negative number [71]. Furthermore, ζ is, by construction, independent of the pairing scheme for real eigensolutions since it only depends on the coefficient matrices. When $n = 1$, each coefficient matrix is just a positive real number and ζ in (4.29) reduces to the well-known damping ratio for single-degree-of-freedom systems in Eq. (4.5).

The system damping ratio (4.29) is invariant under any linear coordinate transformation such as the modal transformation. Suppose a damped system is transformed by an equivalence transformation into

$$\mathbf{WMV}\ddot{\mathbf{p}} + \mathbf{WCV}\dot{\mathbf{p}} + \mathbf{WKVp} = \mathbf{0}, \tag{4.30}$$

where $\mathbf{W}$ and $\mathbf{V}$ are invertible $n \times n$ matrices. Since $\operatorname{tr}\left(\mathbf{V}^{-1}\mathbf{CV}\right) = \operatorname{tr}(\mathbf{C})$ and $\det\left(\mathbf{V}^{-1}\mathbf{KV}\right) = \det(\mathbf{K})$ [71], the parameter ζ_p of the transformed system is

$$\begin{aligned} \zeta_p &= \frac{\operatorname{tr}\left\{(\mathbf{WMV})^{-1}\,\mathbf{WCV}\right\}}{2n} \frac{1}{\sqrt[2n]{\det\left\{(\mathbf{WMV})^{-1}\,\mathbf{WKV}\right\}}} \\ &= \frac{\operatorname{tr}\left(\mathbf{V}^{-1}\mathbf{M}^{-1}\mathbf{CV}\right)}{2n} \frac{1}{\sqrt[2n]{\det\left(\mathbf{V}^{-1}\mathbf{M}^{-1}\mathbf{KV}\right)}} \\ &= \frac{\operatorname{tr}\left(\mathbf{M}^{-1}\mathbf{C}\right)}{2n} \frac{1}{\sqrt[2n]{\det\left(\mathbf{M}^{-1}\mathbf{K}\right)}} = \zeta. \end{aligned} \tag{4.31}$$

Thus, the following theorem was proved:

Theorem 4.2 *The system damping ratio ζ is invariant under equivalence transformation. In particular, ζ is invariant under modal transformations.*

Now, it will be proved that the system response in free vibration is oscillatory if $\zeta < 1$. The $2n$ system eigenvalues λ_j are the same as those of the $2n \times 2n$ state companion matrix given by

$$\mathbf{A} = \begin{bmatrix} \mathbf{0} & \mathbf{I} \\ -\mathbf{M}^{-1}\mathbf{K} & -\mathbf{M}^{-1}\mathbf{C} \end{bmatrix}. \tag{4.32}$$

The trace and determinant of $\mathbf{A}$ can both be related to its eigenvalues [71]. As a consequence [95],

$$\prod_{j=1}^{2n} \lambda_j = \lambda_1\lambda_2 \ldots \lambda_{2n} = \det(\mathbf{A}) = \det\left(\mathbf{M}^{-1}\mathbf{K}\right) = \frac{\det(\mathbf{K})}{\det(\mathbf{M})}, \tag{4.33}$$

and

$$\sum_{j=1}^{2n} \lambda_j = \operatorname{tr}(\mathbf{A}) = -\operatorname{tr}\left(\mathbf{M}^{-1}\mathbf{C}\right). \tag{4.34}$$

It follows that

$$\zeta = -\frac{\sum\limits_{j=1}^{2n} \lambda_j}{2n} \frac{1}{\left(\prod\limits_{j=1}^{2n} \lambda_j\right)^{1/2n}} = -\frac{\text{arithmetic mean of eigenvalues}}{\text{geometric mean of eigenvalues}}. \tag{4.35}$$

Again, the terms arithmetic and geometric mean are used loosely since each λ_j ($j = 1, 2, \ldots, 2n$) is either a real negative number or a complex number with negative real part. The arithmetic mean of all $-\lambda_j$ can be written as

$$-\frac{1}{2n}\sum_{j=1}^{2n} \lambda_j = \frac{1}{n}\sum_{j=1}^{n}\left(\frac{-\lambda_j - \lambda_{n+j}}{2}\right), \tag{4.36}$$

where each pair $\{\lambda_j, \lambda_{n+j}\}$ ($j = 1, 2, \ldots, n$) is the same pairing utilized in phase synchronization. Since complex conjugate eigensolutions must be paired in phase synchronization,

$$-\lambda_j - \lambda_{n+j} = 2|\mathrm{Re}(\lambda_j)| \tag{4.37}$$

is always twice the absolute value of the real part of a complex eigenvalue λ_j. Real eigenvalues are negative, so each pairing of real eigenvalues in (4.36) results in a positive real number. Therefore, pairing results in the arithmetic mean of n positive numbers, and the inequality governing arithmetic and geometric means is applicable [85]:

$$\frac{1}{n}\sum_{j=1}^{n}\left(\frac{-\lambda_j-\lambda_{n+j}}{2}\right)\geq\left[\prod_{j=1}^{n}\left(\frac{-\lambda_j-\lambda_{n+j}}{2}\right)\right]^{1/n}. \tag{4.38}$$

It follows that

$$\zeta\geq\left[\prod_{j=1}^{n}\left(\frac{-\lambda_j-\lambda_{n+j}}{2}\right)\right]^{1/n}\frac{1}{\left(\prod_{j=1}^{2n}\lambda_j\right)^{1/2n}}=\left[\prod_{j=1}^{n}\left(\frac{-\lambda_j-\lambda_{n+j}}{2\sqrt{\lambda_j\lambda_{n+j}}}\right)\right]^{1/n}. \tag{4.39}$$

This means

$$\zeta\geq\left(\zeta_{p1}\zeta_{p2}\cdots\zeta_{pn}\right)^{1/n}, \tag{4.40}$$

where ζ_{pj} $(j=1,\ldots,n)$ is given by (4.7). If $\zeta<1$, at least one $\zeta_{pj}<1$. Thus, the following theorem holds:

Theorem 4.3 *The free response of a damped system is oscillatory if the system damping ratio* $0<\zeta<1$.

In other words, at least one degree of freedom is underdamped. Again, the system response is oscillatory if at least one scalar equation in the decoupled system is underdamped due to coupling. It is practical to use Theorem 4.3 first to check for oscillatory behavior and, if necessary, Theorem 4.1 can then be used for a complete characterization.

For single-degree-of-freedom systems, $\zeta>1$ indicates overdamping. For multi-degree-of-freedom systems, this result is no longer valid. If all degrees of freedom are overdamped,

$$\zeta_{pj}>1 \tag{4.41}$$

for all j $(j=1,\ldots,n)$. Because $\zeta\geq\left(\zeta_{p1}\zeta_{p2}\cdots\zeta_{pn}\right)^{1/n}$, this clearly means $\zeta>1$. However, it is not necessary for an overdamped degree of freedom to occur when $\zeta>1$. In fact, all degrees of freedom may be underdamped even when $\zeta>1$.

Theorem 4.4 *If all degrees of freedom are overdamped, then the system damping ratio* $\zeta>1$.

When $\zeta = 1$, there are two mutually exclusive possibilities. First, all λ_j ($j = 1, 2, \ldots, 2n$) are real. Then ζ is a ratio of arithmetic mean to geometric mean of $2n$ real numbers, and $\zeta = 1$ if and only if λ_j are all equal. In this case, all degrees of freedom are identically and critically damped. Second, some λ_j are complex. Complex conjugate eigenvalues combine to decrease ζ, but real and unequal eigenvalues combine to increase ζ. In general, underdamped, overdamped, and critically damped degrees of freedom can occur together when $\zeta = 1$.

While the system damping ratio indicates oscillatory behavior if $0 < \zeta < 1$, no information can be deduced for the cases when $\zeta \geq 1$. That might indicate a limitation when characterizing the response of a large number of equations by a single number.

4.1.2.1 Example 4.3

Consider a system with

$$\mathbf{M} = \mathbf{I}, \quad \mathbf{C} = \begin{bmatrix} 3.4 & 1 \\ 1 & 3 \end{bmatrix}, \quad \mathbf{K} = \begin{bmatrix} 2 & -4 \\ -4 & 26 \end{bmatrix}. \tag{4.42}$$

The system damping ratio is

$$\zeta = \frac{\mathrm{tr}\,(\mathbf{C})}{4} \frac{1}{\sqrt[4]{\det\,(\mathbf{K})}} = 0.6532, \tag{4.43}$$

so at least one degree of freedom is underdamped and oscillatory behavior is present by Theorem 4.3. To see that, arrange the eigensolutions in primary–secondary scheme as

$$\mathbf{\Lambda}_1 = \begin{bmatrix} -1.2957 + 4.9258i & 0 \\ 0 & -3.4005 \end{bmatrix}, \tag{4.44}$$

$$\mathbf{\Lambda}_2 = \begin{bmatrix} -1.2957 - 4.9258i & 0 \\ 0 & -0.4081 \end{bmatrix}. \tag{4.45}$$

By Theorem 4.1, this system has one overdamped and one underdamped degree of freedom. The decoupled system is given by

$$\ddot{\mathbf{p}} + \begin{bmatrix} 2.5914 & 0 \\ 0 & 3.8086 \end{bmatrix} \dot{\mathbf{p}} + \begin{bmatrix} 25.9427 & 0 \\ 0 & 1.3877 \end{bmatrix} \mathbf{p} = \mathbf{0}. \tag{4.46}$$

The system has an underdamped and an overdamped degree of freedom with $\zeta_{p1} = 0.2544$ and $\zeta_{p2} = 1.6165$, respectively.

4.1.2.2 Example 4.4

Consider a two-degree-of-freedom system:

$$\ddot{\mathbf{q}} + \begin{bmatrix} 10 & -4 \\ -4 & 10 \end{bmatrix} \dot{\mathbf{q}} + \begin{bmatrix} 31 & -19 \\ -19 & 31 \end{bmatrix} \mathbf{q} = \mathbf{0}. \tag{4.47}$$

This system is classically damped. Using modal analysis, the equation of motion is decoupled into

$$\ddot{\mathbf{p}} + \begin{bmatrix} 14 & 0 \\ 0 & 6 \end{bmatrix} \dot{\mathbf{p}} + \begin{bmatrix} 50 & 0 \\ 0 & 12 \end{bmatrix} \mathbf{p} = \mathbf{0}. \tag{4.48}$$

All degrees of freedom are underdamped with damping ratios given by

$$\zeta_{p1} = \frac{14}{2\sqrt{50}} = 0.9899, \ \zeta_{p2} = \frac{6}{2\sqrt{12}} = 0.8660. \tag{4.49}$$

On the contrary, the system damping ratio is

$$\zeta = \frac{\mathrm{tr}\,(\mathbf{C})}{4} \frac{1}{\sqrt[4]{\det(\mathbf{K})}} = 1.0103. \tag{4.50}$$

As additional verification, the system eigenvalues are $\lambda_1 = -7 + i$ and $\lambda_2 = -7 + i\sqrt{3}$, which also produces

$$\zeta = -\frac{\lambda_1 + \lambda_2 + \bar{\lambda}_1 + \bar{\lambda}_2}{4} \frac{1}{\sqrt[4]{\lambda_1 \lambda_2 \bar{\lambda}_1 \bar{\lambda}_2}} = 1.0103. \tag{4.51}$$

4.1.2.3 Example 4.5

Consider a damped system with

$$\mathbf{M} = \mathbf{I}, \quad \mathbf{C} = \begin{bmatrix} 2 & -1 \\ -1 & 2 \end{bmatrix}, \quad \mathbf{K} = \begin{bmatrix} 5 & -2 \\ -2 & 1 \end{bmatrix}. \tag{4.52}$$

Solution of the quadratic eigenvalue problem yields

$$\mathbf{\Lambda}_1 = \begin{bmatrix} \lambda_1 & 0 \\ 0 & \lambda_2 \end{bmatrix} = \begin{bmatrix} -1.2608 + 1.8669i & 0 \\ 0 & -1.3303 \end{bmatrix}, \tag{4.53}$$

and

$$\mathbf{\Lambda}_2 = \begin{bmatrix} \lambda_3 & 0 \\ 0 & \lambda_4 \end{bmatrix} = \begin{bmatrix} -1.2608 - 1.8669i & 0 \\ 0 & -0.1481 \end{bmatrix}. \tag{4.54}$$

This system has an underdamped and an overdamped degree of freedom with $\zeta_{p1} = 0.5597$ and $\zeta_{p2} = 1.6653$, respectively, while $\zeta = 1$.

4.2 A Canonical Form of the Equation of Motion of Linear Systems

It is well known that damped systems either of a single degree or under classical damping can be reduced to an undamped form. For example, a single-degree-of-freedom system

$$m\ddot{q} + c\dot{q} + kq = f(t) \tag{4.55}$$

is converted into

$$\ddot{x} + \left(\frac{k}{m} - \frac{c^2}{4m^2}\right)x = \frac{1}{m}\exp\left(\frac{c}{2m}t\right)f(t) \tag{4.56}$$

using the invertible transformation [3, 96–98]

$$x = \exp\left(\frac{c}{2m}t\right)q. \tag{4.57}$$

This undamped form is sometimes referred to as the normal form. Perhaps it would not be surprising that transformation to an undamped form involves an exponential factor. In free vibration, the response $q(t)$ decays exponentially with any amount of viscous damping. This decay is arrested by the exponential term in Eq. (4.57), which also exponentially magnifies the excitation of $x(t)$ in Eq. (4.56).

If a damped system is classically damped, then it can be decoupled by modal analysis into a series of independent single-degree-of-freedom systems. A classically damped system $\mathbf{M}\ddot{\mathbf{q}} + \mathbf{C}\dot{\mathbf{q}} + \mathbf{K}\mathbf{q} = \mathbf{f}(t)$ can be converted into $\ddot{\mathbf{p}} + \mathbf{D}\dot{\mathbf{p}} + \boldsymbol{\Omega}\mathbf{p} = \mathbf{g}(t)$ for which the modal damping matrix $\mathbf{D} = \mathbf{U}^T\mathbf{C}\mathbf{U}$ is diagonal. To eliminate the damping term in modal coordinates, apply the transformation:

$$\mathbf{x} = \exp\left(\frac{1}{2}\mathbf{D}t\right)\mathbf{p}. \tag{4.58}$$

Observe that $\exp(\mathbf{D}t/2)$ is a diagonal matrix. Upon transformation, the equation of motion in modal coordinates is converted into

$$\ddot{\mathbf{x}} + \left(\boldsymbol{\Omega} - \frac{1}{4}\mathbf{D}^2\right)\mathbf{x} = \exp\left(\frac{\mathbf{D}t}{2}\right)\mathbf{U}^T\mathbf{f}(t). \tag{4.59}$$

The original system becomes undamped and decoupled with respect to the coordinate $\mathbf{x}$, which is connected with $\mathbf{q}$ by

$$\mathbf{q} = \mathbf{U}\exp\left(-\frac{1}{2}\mathbf{D}t\right)\mathbf{x}. \tag{4.60}$$

The key to successful reduction of the equation of motion of classically damped linear systems, as described earlier, is decoupling in real space. Therefore, phase synchronization can be used to reduce the equation of motion of a linear system, through an invertible transformation, to the real decoupled form:

$$\ddot{\mathbf{x}} + \mathbf{R}\mathbf{x} = \mathbf{h}(t), \tag{4.61}$$

where $\mathbf{R}$ is a diagonal matrix, and the generalized displacement $\mathbf{x}$ and excitation $\mathbf{h}(t)$ are n-dimensional column vectors. This reduction is applicable to both non-defective and defective linear systems possessing either symmetric or non-symmetric coefficient matrices, meaning that a damped system can always be reduced to an undamped form [82]. As with phase synchronization, all parameters required to construct the invertible transformation are obtained through the solution of the quadratic eigenvalue problem (1.21). In the reduction of the equation of motion, the canonical form specified by Eq. (4.61) is the simplest representation of linear systems.

4.2.1 Systems Possessing Distinct Eigenvalues

When the eigenvalues of Eq. (1.21) are distinct, phase synchronization decouples $\mathbf{A}_2\ddot{\mathbf{q}} + \mathbf{A}_1\dot{\mathbf{q}} + \mathbf{A}_0\mathbf{q} = \mathbf{f}(t)$ into $\ddot{\mathbf{p}} + \mathbf{D}\dot{\mathbf{p}} + \boldsymbol{\Omega}\mathbf{p} = \mathbf{g}(t)$ by either the configuration-space transformation (3.17) or the state-space transformation (3.18). To eliminate the velocity term in the decoupled equation, apply the transformation

$$\mathbf{x} = \exp\left(\frac{1}{2}\mathbf{D}t\right)\mathbf{p}. \tag{4.62}$$

Upon transformation, the decoupled equation is converted into Eq. (4.61), for which

$$\mathbf{R} = \boldsymbol{\Omega} - \frac{1}{4}\mathbf{D}^2 \tag{4.63}$$

is a real diagonal matrix, and

$$\mathbf{h}(t) = \exp\left(\frac{1}{2}\mathbf{D}t\right)\mathbf{g}(t) = \exp\left(\frac{1}{2}\mathbf{D}t\right)\left\{\left(\mathbf{D} + \mathbf{I}\frac{d}{dt}\right)\mathbf{G}_1\mathbf{f}(t) + \mathbf{G}_2\mathbf{f}(t)\right\}. \tag{4.64}$$

Consequently, when recast in the generalized coordinate $\mathbf{x}$, the equation of motion takes on a decoupled form devoid of the velocity term. To determine a configuration-space transformation between $\mathbf{q}$ and $\mathbf{x}$, combine Eqs. (3.17) and (4.62) to yield

$$\mathbf{q} = \left(\mathbf{T}_1 + \mathbf{T}_2\frac{d}{dt}\right)\exp\left(-\frac{1}{2}\mathbf{D}t\right)\mathbf{x} - \mathbf{T}_2\mathbf{G}_1\mathbf{f}(t). \tag{4.65}$$

Alternatively, a state-space transformation can be determined by combining Eqs. (3.18) and (4.62) to obtain

$$\begin{bmatrix}\mathbf{x}(t)\\ \dot{\mathbf{x}}(t)\end{bmatrix} = \begin{bmatrix}\exp\left(\frac{1}{2}\mathbf{D}t\right) & \mathbf{0}\\ \frac{1}{2}\mathbf{D}\exp\left(\frac{1}{2}\mathbf{D}t\right) & \exp\left(\frac{1}{2}\mathbf{D}t\right)\end{bmatrix}\left\{\mathbf{S}\begin{bmatrix}\mathbf{q}(t)\\ \dot{\mathbf{q}}(t)\end{bmatrix} + \begin{bmatrix}\mathbf{0}\\ \mathbf{G}_1\mathbf{f}(t)\end{bmatrix}\right\}. \tag{4.66}$$

When $t = 0$, the above state-space transformation generates the initial values $\mathbf{x}(0)$ and $\dot{\mathbf{x}}(0)$ of the canonical form (4.61). The transformations given by Eqs. (4.65) and (4.66) are real, non-linear, and invertible. In the reduction of an equation of motion, the canonical form specified by (4.61) is the simplest representation that one may achieve.

How many different canonical forms, of the type defined by (4.61) into which an equation of motion can be reduced, are there? It is obvious that the canonical form (4.61) is unique if phase synchronization generates a unique decoupled system. However, phase synchronization can be implemented with different indexing and normalization schemes. For a given indexing scheme, the coefficient matrices $\mathbf{D}$ and $\mathbf{\Omega}$ are independent of the normalization of eigenvectors. As a result, the homogeneous part of Eq. (4.61) remains unchanged by normalization. By contrast, the excitation $\mathbf{h}(t)$ of Eq. (4.61) is dependent on the eigenvectors and, therefore, on the normalization used. For a given indexing scheme, the canonical form (4.61) is unique up to the normalization of eigenvectors.

There remains the question of equivalence due to different indexing schemes. Indexing schemes that re-order the component equations of (4.61) are considered equivalent. Any indexing scheme must pair the complex conjugate eigensolutions. For a given normalization scheme, there is only one decoupled system if all eigenvalues are complex, and, therefore, only one canonical form defined by (4.61). If there are $2n_r$ distinct real eigenvalues, then there are $N = (2n_r)!/(2^{n_r}n_r!)$ different ways to pair the real eigensolutions. Indeed, using a fixed normalization but different indexing schemes, there are N different decoupled systems, and hence N different canonical forms defined by Eq. (4.61). These N canonical forms usually have different homogeneous parts. For a non-defective system with repeated eigenvalues, the number of different canonical forms is less than N. It can be stated that various indexing and normalization schemes generate an equivalence class of canonical forms of the type defined by equation (4.61). However, there are not more than N members of this equivalence class that are essentially different with different homogeneous parts.

4.2.1.1 Example 4.6

Consider a two-degree-of-freedom system governed by

$$\ddot{\mathbf{q}} + \begin{bmatrix} 0.1 & 0.2 \\ 0.1 & 0.3 \end{bmatrix} \dot{\mathbf{q}} + \begin{bmatrix} 0.7 & 0.3 \\ 0.5 & 0.4 \end{bmatrix} \mathbf{q} = \begin{bmatrix} 1 \\ -1 \end{bmatrix} \sin(2t), \tag{4.67}$$

with initial values $\mathbf{q}(0) = \mathbf{0}$ and $\dot{\mathbf{q}}(0) = \mathbf{0}$. Solution of the quadratic eigenvalue problem yields

$$\mathbf{\Lambda}_1 = \begin{bmatrix} -0.0402 + 0.3683i & 0 \\ 0 & -0.1598 + 0.9599i \end{bmatrix}, \quad \mathbf{\Lambda}_2 = \bar{\mathbf{\Lambda}}_1 \tag{4.68}$$

and normalized eigenvectors[1]

$$\mathbf{V}_1 = \begin{bmatrix} 0.4756 + 0.1059i & 0.7404 - 0.0497i \\ -0.9092 + 0.0139i & 0.6698 + 0.0632i \end{bmatrix}, \quad \mathbf{V}_2 = \overline{\mathbf{V}}_1. \tag{4.69}$$

Since all eigenvalues are complex and distinct, there is only one canonical form of the type defined by Eq. (4.61), unique up to the normalization of eigenvectors. The real and diagonal coefficients of the decoupled system are given by

$$\mathbf{D} = \begin{bmatrix} 0.0804 & 0 \\ 0 & 0.3196 \end{bmatrix}, \quad \mathbf{\Omega} = \begin{bmatrix} 0.1373 & 0 \\ 0 & 0.9470 \end{bmatrix}. \tag{4.70}$$

From Eqs. (3.13) and (3.14),

$$\mathbf{G}_1 = \begin{bmatrix} -0.0374 & 0.1698 \\ -0.1770 & 0.2236 \end{bmatrix}, \quad \mathbf{G}_2 = \begin{bmatrix} 0.6808 & -0.7721 \\ 0.9234 & 0.4728 \end{bmatrix}. \tag{4.71}$$

It can be verified that the canonical form (4.61) is specified by

$$\ddot{\mathbf{x}} + \begin{bmatrix} 0.1357 & 0 \\ 0 & 0.9215 \end{bmatrix} \mathbf{x} = \mathbf{h}(t), \tag{4.72}$$

for which

$$\mathbf{h}(t) = \begin{bmatrix} \{-0.4144\cos(2t) + 1.4362\sin(2t)\}e^{0.0402t} \\ \{-0.8012\cos(2t) + 0.3226\sin(2t)\}e^{0.1598t} \end{bmatrix}. \tag{4.73}$$

The forcing terms are depicted in Fig. 4.2.

[1] Eigenvectors have been normalized in accordance with the scheme depicted in Appendix B.

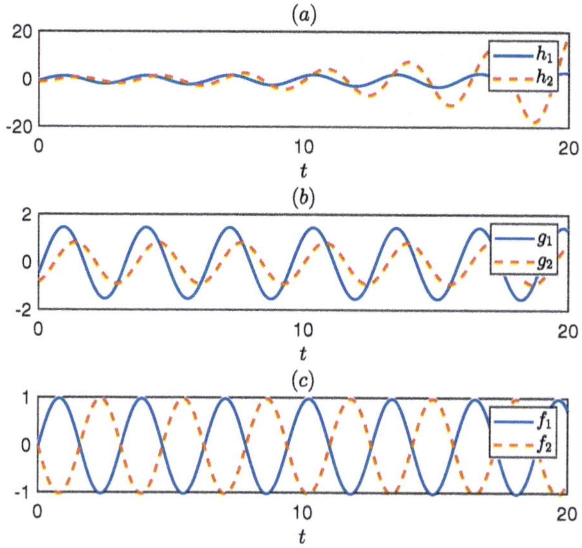

Fig. 4.2 (**a**) Forcing $\mathbf{h}(t)$ of system in canonical form, (**b**) forcing $\mathbf{g}(t)$ of decoupled system and (**c**) $\mathbf{f}(t)$ of Example 4.6. Exponential growth of $\mathbf{h}(t)$ is due to the exponential terms in (4.64)

Using Eq. (4.65), the configuration-space transformation between $\mathbf{q}$ and $\mathbf{x}$ can be expressed as

$$\mathbf{q} = \left(\mathbf{E}(t) + \mathbf{B}(t)\frac{d}{dt}\right)\mathbf{x} + \begin{bmatrix} 0.0388 \\ 0.0342 \end{bmatrix} \sin(2t), \tag{4.74}$$

where

$$\mathbf{E}(t) = \begin{bmatrix} 0.4756e^{-0.0402t} & 0.7404e^{-0.1598t} \\ -0.9092e^{-0.0402t} & 0.6698e^{-0.1598t} \end{bmatrix} \tag{4.75}$$

and

$$\mathbf{B}(t) = \begin{bmatrix} 0.2874e^{-0.0402t} & -0.0518e^{-0.1598t} \\ 0.0377e^{-0.0402t} & 0.0658e^{-0.1598t} \end{bmatrix}. \tag{4.76}$$

The state-space transformation that reduces Eqs. (4.67)–(4.72) is given by Eq. (4.66), for which

$$
\begin{bmatrix} \exp\left(\frac{1}{2}\mathbf{D}t\right) & \mathbf{0} \\ \frac{1}{2}\mathbf{D}\exp\left(\frac{1}{2}\mathbf{D}t\right) & \exp\left(\frac{1}{2}\mathbf{D}t\right) \end{bmatrix}
= \begin{bmatrix} e^{0.0402t} & 0 & 0 & 0 \\ 0 & e^{0.1598t} & 0 & 0 \\ 0.0402e^{0.0402t} & 0 & e^{0.0402t} & 0 \\ 0 & 0.1598e^{0.1598t} & 0 & e^{0.1598t} \end{bmatrix}, \tag{4.77}
$$

$$
\mathbf{S} = \begin{bmatrix} 0.6941 & -0.7286 & -0.0374 & 0.1698 \\ 0.9281 & 0.5045 & -0.1770 & 0.2236 \\ -0.0587 & -0.0567 & 0.6808 & -0.7721 \\ 0.0121 & -0.0363 & 0.9234 & 0.4728 \end{bmatrix} \tag{4.78}
$$

and

$$
\begin{bmatrix} \mathbf{0} \\ \mathbf{G}_1\mathbf{f}(t) \end{bmatrix} = \begin{bmatrix} 0 \\ 0 \\ -0.2072\sin(2t) \\ -0.4006\sin(2t) \end{bmatrix}. \tag{4.79}
$$

The initial values of Eq. (4.72) are $\mathbf{x}(0) = \mathbf{0}$ and $\dot{\mathbf{x}}(0) = \mathbf{0}$. The responses are shown in Fig. 4.3.

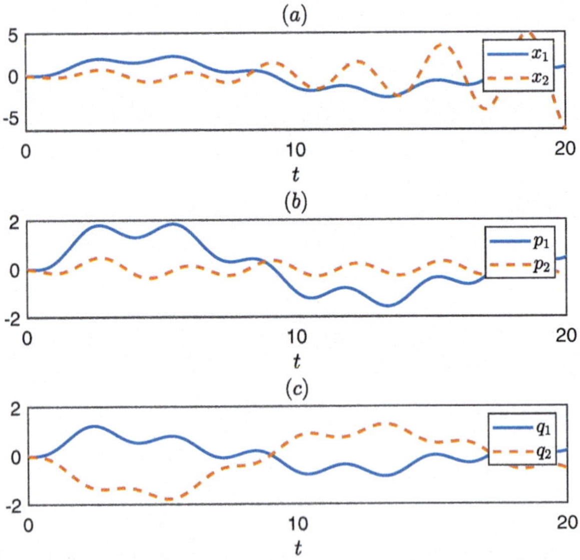

Fig. 4.3 (**a**) Response $\mathbf{x}(t)$ of system in canonical form, (**b**) response $\mathbf{p}(t)$ of decoupled system and (**c**) response $\mathbf{q}(t)$ of coupled system of Example 4.6. The response $\mathbf{x}(t)$ of the canonical system (4.72) grows exponentially due to the excitation $\mathbf{h}(t)$ in Eq. (4.73)

To examine the effect of normalization, let the eigenvectors $\mathbf{v}_j$ $(j = 1, 2)$ be normalized in such a way that the state eigenvectors $[\mathbf{v}_j \quad \lambda_j \mathbf{v}_j]^T$ have unit Euclidean norm. In this case,

$$\mathbf{V}_1 = \begin{bmatrix} 0.4308 + 0.1028i & 0.5309 \\ -0.8265 & 0.4751 + 0.0773i \end{bmatrix}, \quad \mathbf{V}_2 = \overline{\mathbf{V}}_1. \tag{4.80}$$

The homogeneous part of Eq. (4.72) remains unchanged because it is constructed from the eigenvalues. However, the excitation $\mathbf{h}(t)$ in Eq. (4.72) becomes

$$\mathbf{h}(t) = \begin{bmatrix} \{-0.5877 \cos(2t) + 1.5759 \sin(2t)\} e^{0.0402t} \\ \{-1.1927 \cos(2t) + 0.4078 \sin(2t)\} e^{0.1598t} \end{bmatrix}. \tag{4.81}$$

The transformation given either by Eq. (4.65) or Eq. (4.66) also changes with normalization in such a way that Eq. (4.72) with $\mathbf{h}(t)$ specified by Eq. (4.81) is generated. As explained earlier, canonical forms generated by different normalization schemes are regarded as equivalent.

4.2.2 Systems Possessing Repeated or Defective Eigenvalues

After a defective system has been converted into a decoupled system by phase synchronization, the canonical form (4.61) is obtained through application of transformation (4.62). In this case, the diagonal coefficient matrix $\mathbf{R}$ is still given by Eq. (4.63), and the excitation has the form

$$\begin{aligned} \mathbf{h}(t) &= \exp\left(\frac{1}{2}\mathbf{D}t\right)\mathbf{g}(t) \\ &= \exp\left(\frac{1}{2}\mathbf{D}t\right)\left\{\left(\mathbf{D} + \mathbf{I}\frac{d}{dt}\right)\mathbf{G}_1(t)\mathbf{f}(t) + \mathbf{G}_2(t)\mathbf{f}(t)\right\}. \end{aligned} \tag{4.82}$$

Combining (4.62) and (3.76) yields the configuration-space transformation relating $\mathbf{q}$ and $\mathbf{x}$:

$$\mathbf{q} = \left(\mathbf{T}_1(t) + \mathbf{T}_2(t)\frac{d}{dt}\right)\exp\left(-\frac{1}{2}\mathbf{D}t\right)\mathbf{x} - \mathbf{T}_2(t)\mathbf{G}_1(t)\mathbf{f}(t). \tag{4.83}$$

When (4.62) and (3.75) are combined, the transformation connecting $\mathbf{q}$ and $\mathbf{x}$ in the state space is obtained:

$$\begin{aligned} \begin{bmatrix} \mathbf{x}(t) \\ \dot{\mathbf{x}}(t) \end{bmatrix} = &\begin{bmatrix} \exp\left(\frac{1}{2}\mathbf{D}t\right) & \mathbf{0} \\ \frac{1}{2}\mathbf{D}\exp\left(\frac{1}{2}\mathbf{D}t\right) & \exp\left(\frac{1}{2}\mathbf{D}t\right) \end{bmatrix} \\ &\left\{\mathbf{T}^{-1}(t)\begin{bmatrix} \mathbf{q}(t) \\ \dot{\mathbf{q}}(t) \end{bmatrix} + \begin{bmatrix} \mathbf{0} \\ \mathbf{G}_1(t)\mathbf{f}(t) \end{bmatrix}\right\}. \end{aligned} \tag{4.84}$$

Should the eigenvalues be distinct, then it is easy to verify that all formulae for transforming a linear system into the canonical form (4.61) reduce to their counterparts in Sect. 4.2.1.

4.2.2.1 Example 4.7

Consider the defective system of Example 3.5. The canonical form (4.61) for this system is specified by

$$\ddot{\mathbf{x}} + \begin{bmatrix} 6 & 0 \\ 0 & 6 \end{bmatrix} \mathbf{x} = \mathbf{h}(t), \tag{4.85}$$

where the excitation

$$\mathbf{h}(t) = \frac{e^t}{6} \begin{bmatrix} -8\cos(t) - t\{\cos(t) + 2\sin(t)\} \\ \cos(t) + 2\sin(t) \end{bmatrix}. \tag{4.86}$$

The coordinates $\mathbf{q}$ and $\mathbf{x}$ are related in the configuration space by transformation (4.83):

$$\mathbf{q} = \left(\begin{bmatrix} 0 & 5e^{-t}/2 \\ e^{-t} & te^{-t} \end{bmatrix} + \begin{bmatrix} -e^{-t}/2 & -te^{-t}/2 \\ 0 & 0 \end{bmatrix} \frac{d}{dt} \right) \mathbf{x}. \tag{4.87}$$

Reduction of the coupled system to Eq. (4.85) is accomplished in state space by Eq. (4.84), for which

$$\begin{bmatrix} \exp\left(\frac{1}{2}\mathbf{D}t\right) & \mathbf{0} \\ \frac{1}{2}\mathbf{D}\exp\left(\frac{1}{2}\mathbf{D}t\right) & \exp\left(\frac{1}{2}\mathbf{D}t\right) \end{bmatrix} = \begin{bmatrix} 1 & 0 & 0 & 0 \\ 0 & 1 & 0 & 0 \\ 1 & 0 & 1 & 0 \\ 0 & 1 & 0 & 1 \end{bmatrix} e^t, \tag{4.88}$$

and

$$\begin{bmatrix} \mathbf{0} \\ \mathbf{G}_1(t)\mathbf{f}(t) \end{bmatrix} = \frac{1}{3} \begin{bmatrix} 0 \\ 0 \\ t\cos(t) \\ -\cos(t) \end{bmatrix}. \tag{4.89}$$

The initial values of Eq. (4.85) computed from (4.84) are $\mathbf{x} = \mathbf{0}$ and $\dot{\mathbf{x}} = [0 \quad -1/3]^T$. The homogeneous part of Eq. (4.85) is unaffected by normalization because it is constructed only from the eigenvalues. The transformation given either (4.83) or (4.84) changes with normalization. As in the non-defective case, the canonical form generated is dependent on the normalization of eigenvectors through $\mathbf{h}(t)$.

4.3 Earthquake Engineering and Modal Reduction

The linearized equation of motion (1.10) is widely used in earthquake engineering to model the dynamic behavior of, for example, multistory buildings, nuclear power plants, or base-isolated structures [66–69, 99].

Consider the idealized 10-story building in Fig. 4.4 subject to ground motion q_g, here modeled by the 1940 El Centro earthquake (see Chapter 6 of [4] for data on the earthquake). The motion of the building is described by Chopra [4], Humar [6], and Itoh [69]:

$$\mathbf{M}\ddot{\mathbf{q}} + \mathbf{C}\dot{\mathbf{q}} + \mathbf{K}\mathbf{q} = -\mathbf{M}\mathbf{1}\ddot{q}_g, \tag{4.90}$$

where $\mathbf{1}$ is a 10-dimensional column vector of 1s. All three coefficient matrices are symmetric and positive-definite.

Engineers may be interested in the issuing motion when the building is excited by an earthquake q_g. In addition, it is desirable to obtain appropriate accuracy in computation while solving fewer differential equations, a procedure often termed model reduction. System (4.90) can be decoupled by phase synchronization. The

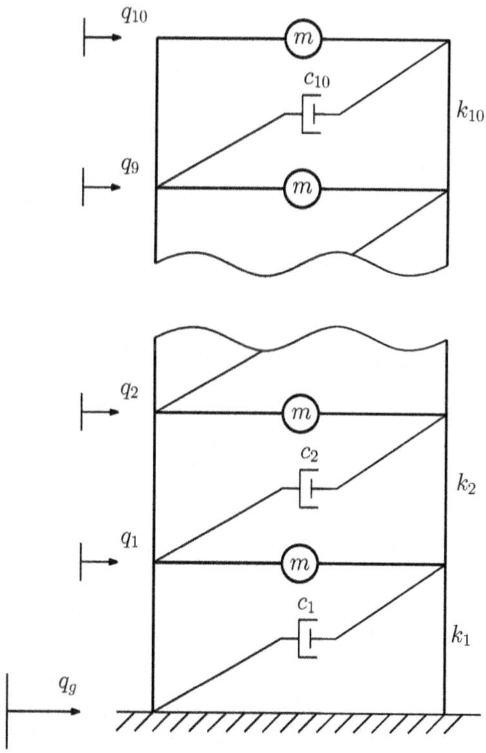

Fig. 4.4 A fictitious 10-story building. All floor levels have the same mass $m = 1$. The stiffness of each floor beam is: $k_1 = 0.06$, $k_2 = 0.18$, $k_3 = 0.12$, $k_4 = 0.06$, $k_5 = 0.19$, $k_6 = 0.09$, $k_7 = 0.14$, $k_8 = 0.15$, $k_9 = 0.08$, $k_{10} = 0.20$. Energy dissipation is modeled by viscous damping, for which $c_1 = 0.10$, $c_2 = 0.09$, $c_3 = 0.11$, $c_4 = 0.09$, $c_5 = 0.17$, $c_6 = 0.10$, $c_7 = 0.06$, $c_8 = 0.14$, $c_9 = 0.10$, $c_{10} = 0.10$. If energy dissipation is associated with the deformation of each story, the viscous dampers may be visualized as shown [4]. Each coordinate q_j $(j = 1, 2, \ldots, 10)$ is measured with respect to the ground motion q_g

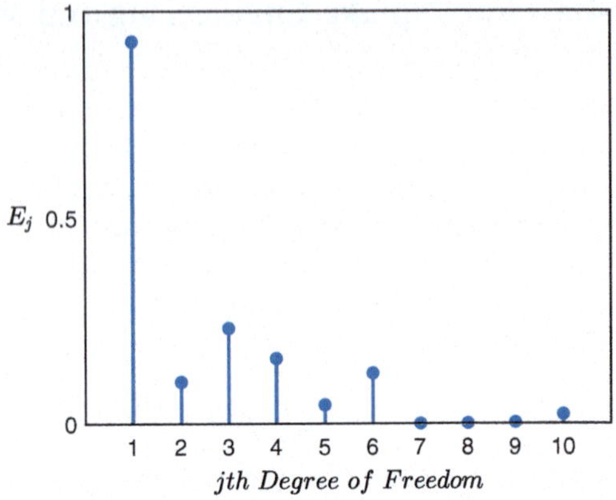

Fig. 4.5 Energy contribution of each decoupled degree of freedom p_j ($j = 1, 2, \ldots, 10$). The greatest contributions are due to the first six degrees of freedom

contribution of each coordinate $p_j(t)$ ($j = 1, 2, \ldots, 10$) can be estimated by computation of the energy [30]:

$$E_j = \frac{1}{2}\int_0^T \left[\dot{p}_j(t) + \Omega_{jj}p_j(t)\right]dt, \tag{4.91}$$

where T is the total time of simulation and Ω_{jj} is the j-th diagonal element of $\mathbf{\Omega}$, as obtained by phase synchronization. The results are shown in Fig. 4.5, and they indicate that an approximation to the response can be obtained by solving only the first six decoupled equations (which are associated with the lowest frequencies) because they carry most of the energy within the system. The approximate response is then obtained by $\mathbf{q} = \mathbf{T}_1\mathbf{p} + \mathbf{T}_2\dot{\mathbf{p}}$.

The classical engineering approach is based on solving the symmetric eigenvalue problem $\mathbf{Ku} = \omega^2\mathbf{M}$ to obtain the mode shapes and natural frequencies of the structure. Upon the modal transformation $\mathbf{q} = \mathbf{Up}$, both mass and stiffness matrices have been diagonalized. Unfortunately, the modal damping matrix $\mathbf{D} = \mathbf{U}^T\mathbf{CU}$ is not diagonalized by the modal transformation because $\mathbf{CM}^{-1}\mathbf{K} \neq \mathbf{KM}^{-1}\mathbf{C}$, meaning each floor has a different level of damping. In this case, it is relatively routine in structural engineering to diagonalize $\mathbf{D}$ by simply neglecting its off-diagonal elements through a procedure termed the decoupling approximation.

This approximation technique may be attributed to Lord Rayleigh. In Section 102 of "The Theory of Sound" in 1894, Lord Rayleigh [1] suggested that the off-diagonal elements of the modal damping matrix may be neglected if they are small in magnitude. This simple procedure is a member of the class of approximate techniques seeking to decouple a damped system by modifying its coefficient matrices [100–104]. Theoretically speaking, it means that coupling as represented by the off-diagonal elements of $\mathbf{D}$ is a second-order effect. This approximation is intuitively appealing if $\mathbf{D}$ is diagonally dominant or if the natural frequencies are well separated when damping is small. An index of diagonality can be defined as [30]

$$\rho(\mathbf{D}) = \frac{\sum_{j=1}^{n} |d_{jj}|}{\sum_{\substack{l,j=1 \\ l \neq j}}^{n} |d_{lj}|}. \tag{4.92}$$

If a modal damping matrix $\mathbf{D}$ is diagonally dominant, then $\rho(\mathbf{D}) \geq 1$ and, for diagonal $\mathbf{D}$, $\rho(\mathbf{D}) = \infty$. In the case of this 10-story building, it reveals that $\mathbf{D}$ is indeed diagonally dominant: $\rho(\mathbf{D}) = 1.4866$. Therefore, one can neglect the off-diagonal elements of $\mathbf{D}$ and then use the first six coordinates of this decoupled system to obtain an approximate response.

The motion of each floor q_j $(j = 1, 2, \ldots, 10)$ is computed by numerical integration, and these are regarded as the building response for comparison purposes. Figure 4.6 shows a comparison of the issuing motion of each floor computed by two different methods: the approximate solution obtained by phase synchronization and the solution obtained by decoupling approximation. Fig. 4.6 indicates that an approximate solution computed by phase synchronization is comparable to the building response, while the response obtained by decoupling approximation can considerably underestimate or overestimate the response. This happens because decoupling approximation is based on neglecting the coupling caused by nonzero off-diagonal elements of the modal damping matrix, while phase synchronization generates a decoupled equation by incorporating all coupling effects properly [30].

4.4 Invariant of Motion for Systems in Free Vibration

An invariant of motion $I(\mathbf{q}, \dot{\mathbf{q}}, t)$ is a quantity that satisfies $\dot{I}(\mathbf{q}, \dot{\mathbf{q}}, t) \equiv 0$ during the motion of the system. Besides theoretical interest, these may render useful checks on numerical simulations for large systems, particularly those that do not conserve energy. Sometimes the Lagrangian $L(\mathbf{q}, \dot{\mathbf{q}}, t)$ can be time dependent [105, 106], so the Jacobi integral cannot be used to find an invariant of motion [107]. However, decoupling of the equations of motion can directly yield conserved quantities.

A Lagrangian is a scalar function $L(\mathbf{q}, \dot{\mathbf{q}}, t)$ that satisfies the corresponding Euler–Lagrange equation:

$$\frac{d}{dt}\left[\frac{\partial L}{\partial \dot{\mathbf{q}}}\right] - \frac{\partial L}{\partial \mathbf{q}} = \mathbf{0}, \tag{4.93}$$

where the derivative of a multivariate scalar function F with respect to an n-dimensional vector such as $\mathbf{q}$ in Eq. (4.93) is defined by

$$\frac{\partial F}{\partial \mathbf{q}} = \begin{bmatrix} \frac{\partial F}{\partial q_1} & \frac{\partial F}{\partial q_2} & \cdots & \frac{\partial F}{\partial q_n} \end{bmatrix}^T. \tag{4.94}$$

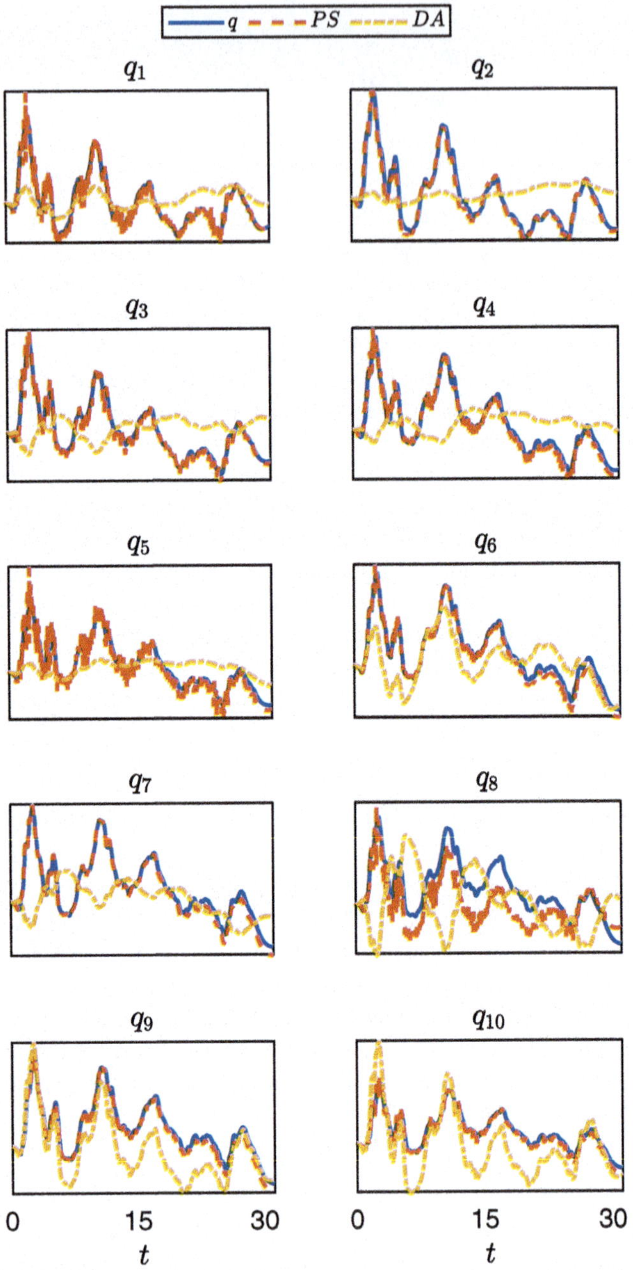

Fig. 4.6 Response of each floor q_j ($j = 1, 2, \ldots, 10$) is computed numerically for comparison purposes. The curves labeled as PS represent the approximate solution obtained by phase synchronization. The curves labeled as DS represent the solution obtained by decoupling approximation. The equations are integrated numerically using `ode45` in MATLAB.

4.4.1 Invariant for Linear Systems

It can be checked that a single-degree-of-freedom system of the form

$$\ddot{p} - (\lambda_1 + \lambda_2)\dot{p} + \lambda_1\lambda_2 p = 0 \tag{4.95}$$

conserves the quantity [105]:

$$I(p, \dot{p}, t) = \frac{1}{2}\left[\dot{p}^2 - (\lambda_1 + \lambda_2)\dot{p}p + \lambda_1\lambda_2 p^2\right] e^{-(\lambda_1+\lambda_2)t}. \tag{4.96}$$

Recall that the decoupled system obtained by phase synchronization $\ddot{\mathbf{p}}+\mathbf{D}\dot{\mathbf{p}}+\boldsymbol{\Omega}\mathbf{p} = \mathbf{0}$ has component equations of the same form as in (4.95). Inspired by Eq. (4.96), this decoupled system admits an invariant of the form

$$\begin{aligned} I(\mathbf{p}, \dot{\mathbf{p}}, t) &= \sum_{j=1}^{n} \frac{1}{2}\left[\dot{p}_j^2 - (\lambda_j + \lambda_{n+j})\dot{p}_j p_j + \lambda_j\lambda_{n+j} p_j^2\right] e^{-(\lambda_j+\lambda_{n+j})t} \\ &= \frac{1}{2}\left(\dot{\mathbf{p}}^T e^{\mathbf{D}t}\dot{\mathbf{p}} + \dot{\mathbf{p}}^T e^{\mathbf{D}t}\mathbf{D}\mathbf{p} + \mathbf{p}^T e^{\mathbf{D}t}\boldsymbol{\Omega}\mathbf{p}\right). \end{aligned} \tag{4.97}$$

Since the decoupling transformation in state space is expressible as

$$\begin{bmatrix}\mathbf{p}\\ \dot{\mathbf{p}}\end{bmatrix} = \mathbf{S}(t)\begin{bmatrix}\mathbf{q}\\ \dot{\mathbf{q}}\end{bmatrix} = \begin{bmatrix}\mathbf{S}_1(t) & \mathbf{S}_2(t)\\ \mathbf{S}_3(t) & \mathbf{S}_4(t)\end{bmatrix}\begin{bmatrix}\mathbf{q}\\ \dot{\mathbf{q}}\end{bmatrix}, \tag{4.98}$$

it is possible to use

$$\mathbf{p} = \mathbf{S}_1(t)\mathbf{q} + \mathbf{S}_2(t)\dot{\mathbf{q}} \tag{4.99}$$

$$\dot{\mathbf{p}} = \mathbf{S}_3(t)\mathbf{q} + \mathbf{S}_4(t)\dot{\mathbf{q}} \tag{4.100}$$

to transform coordinates. An invariant of motion for the coupled system $\mathbf{A}_2\ddot{\mathbf{q}} + \mathbf{A}_1\dot{\mathbf{q}} + \mathbf{A}_0\mathbf{q} = \mathbf{0}$ can be expressed as

$$I(\mathbf{q}, \dot{\mathbf{q}}, t) = \frac{1}{2}\mathbf{q}^T\mathbf{B}_1(t)\mathbf{q} + \frac{1}{2}\dot{\mathbf{q}}^T\mathbf{B}_2(t)\dot{\mathbf{q}} + \dot{\mathbf{q}}^T\mathbf{B}_3(t)\mathbf{q} \tag{4.101}$$

where

$$\mathbf{B}_1(t) = \mathbf{S}_3^T e^{\mathbf{D}t}\mathbf{S}_3 + \mathbf{S}_3^T e^{\mathbf{D}t}\mathbf{D}\mathbf{S}_1 + \mathbf{S}_1^T e^{\mathbf{D}t}\boldsymbol{\Omega}\mathbf{S}_1, \tag{4.102}$$

$$\mathbf{B}_2(t) = \mathbf{S}_4^T e^{\mathbf{D}t}\mathbf{S}_4 + \mathbf{S}_4^T e^{\mathbf{D}t}\mathbf{D}\mathbf{S}_2 + \mathbf{S}_2^T e^{\mathbf{D}t}\boldsymbol{\Omega}\mathbf{S}_2, \tag{4.103}$$

$$\mathbf{B}_3(t) = \mathbf{S}_4^T e^{\mathbf{D}t}\mathbf{S}_3 + \frac{1}{2}\left(\mathbf{S}_4^T e^{\mathbf{D}t}\mathbf{D}\mathbf{S}_1 + \mathbf{S}_2^T e^{\mathbf{D}t}\mathbf{D}\mathbf{S}_3\right) + \mathbf{S}_2^T e^{\mathbf{D}t}\boldsymbol{\Omega}\mathbf{S}_1. \tag{4.104}$$

Note that each submatrix $\mathbf{S}_j(t)$ $(j = 1,\ 2,\ 3,\ 4)$ is of dimension $n \times n$.

4.4.1.1 Example 4.8

Consider the defective system of Example 3.4. Mechanical energy

$$E = \frac{1}{2}\dot{\mathbf{q}}^T \mathbf{M}\dot{\mathbf{q}} + \frac{1}{2}\mathbf{q}^T \mathbf{K}\mathbf{q} \tag{4.105}$$

is not conserved because

$$\dot{E} = -\dot{\mathbf{q}}^T \mathbf{C}\dot{\mathbf{q}} < 0. \tag{4.106}$$

However, it conserves a quantity of the form (4.101). Note that

$$\mathbf{S}(t) = \begin{bmatrix} \mathbf{S}_1(t) & \mathbf{S}_2(t) \\ \mathbf{S}_3(t) & \mathbf{S}_4(t) \end{bmatrix}$$
$$= \begin{bmatrix} -t/4 & 1-t/4 & 0 & -t/4 \\ 1/4 & 1/4 & 0 & 1/4 \\ -t/4-1/4 & 5t/4-1/4 & -t/2 & t/4+3/4 \\ 1/4 & -5/4 & 1/2 & -1/4 \end{bmatrix} \tag{4.107}$$

so the invariant of motion is

$$\begin{aligned} I(\mathbf{q}, \dot{\mathbf{q}}, t) = e^{2t} \Bigg[&\left(\frac{7}{32}q_1^2 + \frac{59}{32}q_2^2 + \frac{1}{8}\dot{q}_1^2 + \frac{11}{32}\dot{q}_2^2 - \frac{9}{16}q_1q_2 + \frac{1}{4}q_1\dot{q}_1 \right. \\ &\left. -\frac{1}{16}q_1\dot{q}_2 - \frac{1}{2}q_2\dot{q}_1 + \frac{11}{16}q_2\dot{q}_2 \right) + \left(\frac{1}{8}q_1^2 + \frac{1}{4}q_2^2 - \frac{9}{8}q_1q_2 \right. \\ &\left. +\frac{1}{8}q_1\dot{q}_1 - \frac{3}{8}q_1\dot{q}_2 - \frac{3}{8}q_2\dot{q}_1 + \frac{1}{4}q_2\dot{q}_2 - \frac{3}{8}\dot{q}_1\dot{q}_2 \right) t + \left(\frac{3}{16}q_1^2 \right. \\ &+\frac{9}{16}q_2^2 + \frac{1}{8}\dot{q}_1^2 + \frac{1}{16}\dot{q}_2^2 - \frac{3}{8}q_1q_2 + \frac{1}{4}q_1\dot{q}_1 + \frac{1}{8}q_1\dot{q}_2 - \frac{1}{2}q_2\dot{q}_1 \\ &\left. +\frac{1}{8}q_2\dot{q}_2 \right) t^2 \Bigg]. \end{aligned} \tag{4.108}$$

It can be verified that $\dot{I} \equiv 0$ with $\ddot{q}_1 = \dot{q}_2 - 2\dot{q}_1 - 2q_1 + q_2$ and $\ddot{q}_2 = \dot{q}_1 - 2\dot{q}_2 + q_1 - 5q_2$.

4.4.2 *Simplification for Classically Damped Systems*

In the advent that the coupled system is classically damped, $\mathbf{S}_2 = \mathbf{S}_3 = \mathbf{0}$ and $\mathbf{S}_1 = \mathbf{S}_4 = \mathbf{U}^{-1}$ up to arbitrary signs in the columns of the mass-normalized modal matrix $\mathbf{U}$. Without loss of generality, take $\mathbf{M} = \mathbf{I}$ so $\mathbf{U}^{-1} = \mathbf{U}^T$. Then, the invariant (4.101) simplifies to

$$I(\mathbf{q}, \dot{\mathbf{q}}, t) = \frac{1}{2}\left(\mathbf{q}^T e^{\mathbf{C}t}\mathbf{K}\mathbf{q} + \dot{\mathbf{q}}^T e^{\mathbf{C}t}\dot{\mathbf{q}} + \dot{\mathbf{q}}^T e^{\mathbf{C}t}\mathbf{C}\mathbf{q} \right), \tag{4.109}$$

which is the same form obtained by Udwadia [105] using modal analysis.

Appendix A
Commonly Used Symbols

A_j	Amplitude of c_j
$\mathbf{A}$	$2n \times 2n$ state companion matrix
$\mathbf{A}_0$	$n \times n$ coefficient of displacement
$\mathbf{A}_1$	$n \times n$ coefficient of velocity
$\mathbf{A}_2$	$n \times n$ coefficient of acceleration
$\mathbf{B}$	$n \times n$ real matrix
c_j	j-th coefficient determined from initial conditions
$\mathbf{C}$	$n \times n$ damping matrix
$\mathbf{D}$	Velocity coefficient of the decoupled system/modal damping matrix
$\mathbf{E}$	$n \times n$ real matrix
$\mathbf{f}(t)$	n-dimensional forcing vector
$\mathbf{g}(t)$	Forcing vector of the decoupled system
$\mathbf{G}_1$	$n \times n$ first coefficient of excitation of decoupled system
$\mathbf{G}_2$	$n \times n$ second coefficient of excitation of decoupled system
$\mathbf{h}(t)$	n-dimensional forcing vector of system in canonical form
$I(\mathbf{q}, \dot{\mathbf{q}}, t)$	Invariant of motion
$\mathbf{I}$	Identity matrix or order n
$\mathbf{J}_p$	$2n \times 2n$ Jordan matrix of decoupled system
$\mathbf{J}_q$	$2n \times 2n$ Jordan matrix of coupled system
$\mathbf{K}$	$n \times n$ stiffness matrix
$L(\mathbf{q}, \dot{\mathbf{q}}, t)$	Lagrangian function
$\mathbf{M}$	$n \times n$ mass matrix
n	System dimension ($n \geq 1$)
n_c	Number of complex conjugate pairs of eigenvalues
n_r	Number of pairs of real eigenvalues
N	Number of different ways to decouple a linear system
$\mathbf{p}$	n-dimensional coordinate vector of the decoupled system
q_j	j-th generalized coordinate
$\mathbf{q}$	n-dimensional column vector of generalized coordinates

R. Gonçalves Salsa Junior, F. Ma, *Advances in the Theory of System Decoupling*,
https://doi.org/10.1007/978-3-030-60846-0

$\mathbf{Q}$	λ-matrix
r_{jl}	Amplitude of l-th component of $\mathbf{v}_j$
$\mathbf{R}$	$n \times n$ coefficient of displacement of system in canonical form
$\mathbf{s}_j$	j-th damped mode of vibration
$\mathbf{S}$	$2n \times 2n$ real transformation matrix
$\mathbf{S}_j$	j-th $n \times n$ block matrix of $\mathbf{S}$
t	Time ($t \geq 0$)
$\mathbf{T}$	Inverse of $\mathbf{S}$
$\mathbf{T}_j$	j-th $n \times n$ block matrix of $\mathbf{T}$
$\mathbf{u}$	Mode shape
$\mathbf{U}$	$n \times n$ modal matrix
$\mathbf{v}$	n-dimensional eigenvector associated with λ
$\mathbf{v}_k^l$	l-th generalized eigenvector associated with a partial multiplicity μ_k
$\mathbf{V}$	$n \times n$ non-singular matrix
$\mathbf{V}_A$	$n \times n$ Vandermonde matrix
$\mathbf{V}_p$	$n \times 2n$ matrix of eigenvectors and generalized eigenvectors of $\mathbf{J}_p$
$\mathbf{V}_q$	$n \times 2n$ matrix of eigenvectors and generalized eigenvectors of $\mathbf{J}_q$
$\mathbf{V}_1$	$n \times n$ matrix of eigenvectors associated with $\mathbf{\Lambda}_1$
$\mathbf{V}_2$	$n \times n$ matrix of eigenvectors associated with $\mathbf{\Lambda}_2$
$\mathbf{w}$	n-dimensional eigenvector of the adjoint eigenvalue problem
$\mathbf{W}$	$n \times n$ non-singular matrix
$\mathbf{x}$	n-dimensional vector of generalized coordinates of system in canonical form
$\mathbf{y}_j$	j-th classically damped mode of vibration of synchronized system
$\mathbf{z}_j$	j-th mode shape of synchronized system
$\mathbf{0}$	Zero matrix or order n or n-dimensional zero vector
$\mathbf{1}$	n-dimensional column vector with entries equal to 1

Greek Letters

α	Real part of λ
β	Algebraic multiplicity of λ
γ	Geometric multiplicity of λ
ζ	System damping ratio
η_j	j-th coefficient of Caughey damping expansion
θ_j	Phase angle of c_j
λ	Eigenvalue
$\mathbf{\Lambda}_1$	$n \times n$ diagonal matrix of first group of n eigenvalues
$\mathbf{\Lambda}_2$	$n \times n$ diagonal matrix of second group of n eigenvalues
μ_k	k-th partial multiplicity of λ
$\rho(\mathbf{B})$	Index of diagonality of $\mathbf{B}$
ϕ_{jl}	Phase angle of l-th component of $\mathbf{v}_j$
ψ	Eigenvalue of generalized linear eigenvalue problem
ω	Imaginary part of λ
$\mathbf{\Omega}$	Displacement coefficient of the decoupled system/spectral matrix

Special Symbols

$[\mathbf{B}, \mathbf{E}]$	Commutator product of matrices **B** and **E**
$\mathbb{C}$	Set of complex numbers
$\mathbb{L}(\mathbf{B}, \mathbf{E})$	The Lie algebra generated by the matrices **B** and **E**

Superscripts

$-$	Complex conjugate
$*$	Complex conjugate transpose
$\sim$	Real quadratic conjugation
T	Transpose
$\cdot$	Time differentiation

Appendix B
Linear Decoupling Program

A decoupling program, consisting of two MATLAB files, is provided together with user instructions.

B.1 Source Codes

B.1.1 File 1: `MAIN.m`

```
%% MAIN
% Supply user-defined input for decoupling and run this file

%% Linear Decoupling Program

% This program decouples any user-supplied linear system in
% real space with an extension of modal analysis. To be
% specific, let M, C and K be arbitrary real square
% matrices (can be non-symmetric) of order n with M assumed
% nonsingular. A real and invertible transformation is
% generated to convert
%            M*(q dot dot) + C*(q dot) + K*q = f(t)
% into
%            I*(p dot dot) + D*(p dot) + OM*p = g(t)
% for which D and OM are real and diagonal. There are two files
% in the program: MAIN and SystemDecoupling.

clear      % Delete old variables
close all  % Delete old figures
clc        % Clear command window

%% Enter User-Supplied Input Parameters

% Coefficients M, C and K of linear system to be decoupled:
```


R. Gonçalves Salsa Junior, F. Ma, *Advances in the Theory of System Decoupling*,
https://doi.org/10.1007/978-3-030-60846-0

```
M = eye(2);
C = [0.7 -0.1;-0.1 0.2];
K = [2 -1;-1 2];

% Initial displacement qo and velocity Dqo of user-supplied
% system:
qo = [1;2];
Dqo = [-1;1];

% Duration T and time increment dt of simulation:
T = 20;
dt = 0.01;

% Excitation f(t) of user-supplied system:
t = 0:dt:T; % Do not delete this vector of time points
% An excitation such as f = [cos(t);0] is written as
% f = [cos(t);0*t] so that all components have the same
% length but f = zeros(2,1) is allowed
f = [cos(t);sin(2*t)];

%% Generate Output Parameters

[L1,L2,V1,V2,A,D,OM,G1,G2,T1,T2,S,po,Dpo,p,Dp,g,q] =...
                    SystemDecoupling(M,C,K,dt,t,qo,Dqo,f);

%% Interpret Results

% Either type parameter names in MATLAB command window or
% plot output arrays. For example, compare the decoupled
% and user-supplied systems by using the following commands:

subplot(2,1,1)
plot(t,p(1,:), t,p(2,:),'--', 'LineWidth',1)
xlabel('\itt')
ylabel('\bfp\rm(\itt\rm)')
title('\rmResponse of Decoupled System')
legend(' \itp_{\rm1}', ' \itp_{\rm2}')
subplot(2,1,2)
plot(t,q(1,:), t,q(2,:),'--', 'LineWidth',1)
xlabel('\itt')
ylabel('\bfq\rm(\itt\rm)')
title('\rmResponse of User-Supplied System')
legend(' \itq_{\rm1}', ' \itq_{\rm2}')
figure

subplot(2,1,1)
plot(t,g(1,:), t,g(2,:),'--', 'LineWidth',1)
xlabel('\itt')
ylabel('\bfg\rm(\itt\rm)')
title('\rmExcitation of Decoupled System')
legend(' \itg_{\rm1}', ' \itg_{\rm2}')
subplot(2,1,2)
plot(t,f(1,:), t,f(2,:),'--', 'LineWidth',1)
xlabel('\itt')
```

```
ylabel('\bff\rm(\itt\rm)')
title('\rmExcitation of User-Supplied System')
legend(' \itf_{\rm1}', ' \itf_{\rm2}')

%% Appendix

% List of 8 input parameters
% M   - nxn mass matrix (can be non-symmetric)
% C   - nxn damping and gyroscopic matrix (can be non-symmetric)
% K   - nxn stiffness and circulatory matrix (can be
%       non-symmetric)
% qo  - nx1 initial displacement q(0) of user-supplied system
% Dqo - nx1 initial velocity (q dot)(0) of user-supplied system
% T   - 1x1 duration of simulation
% dt  - 1x1 time increment in simulation
% f   - nx1 excitation in format f = [f1(t);f2(t);...;fn(t)]

% List of 18 output parameters
% q   - nx1 displacement [q1(t);q2(t);...;qn(t)] of user-supplied
%       system where each column of output gives q(t) at time t
% p   - nx1 displacement [p1(t);p2(t);...;pn(t)] of decoupled
%       system where each column of output gives p(t) at time t
% Dp  - nx1 velocity (p dot)(t) of decoupled system
% po  - nx1 initial displacement p(0) of decoupled system
% Dpo - nx1 initial velocity (p dot)(0) of decoupled system
% A   - 2nx2n state companion matrix of user-supplied system
% D   - nxn diagonal damping matrix of decoupled system
% OM  - nxn diagonal stiffness matrix of decoupled system
% g   - nx1 excitation [g1(t);g2(t);...;gn(t)] of decoupled
%       system
% G1  - nxn first coefficient of excitation of decoupled system
% G2  - nxn second coefficient of excitation of decoupled system
% T1  - nxn first coefficient of p-q decoupling transformation
% T2  - nxn second coefficient of p-q decoupling transformation
% S   - 2nx2n coefficient of state decoupling transformation
% L1  - nxn diagonal matrix of first group of n eigenvalues
% L2  - nxn diagonal matrix of second group of n eigenvalues
% V1  - nxn matrix of eigenvectors associated with L1
% V2  - nxn matrix of eigenvectors associated with L2
```

B.1.2 File 2: `SystemDecoupling.m`

```
%% SystemDecoupling
% This function file decouples the user-defined system in MAIN

%% Algorithm

% The decoupling theory is expounded in Chapters 2 and 3 of
```

```
7 % the book. A flowchart for the decoupling algorithm is provided
8 % in Fig. 3.2.
9
10 % The experienced user can change the default settings:
11 % 1. On line 140, eigenvectors are not normalized if
12 % switch_normalize = false.
13 % 2. On line 260, default tolerance for determining if some
14 % eigenvalues are numerically close can be changed.
15 % 3. On line 324, primary-secondary pairing of eigensolutions
16 % is not used if switch_primary_secondary = false. A different
17 % pairing scheme can be specified on line 331 by changing the
18 % indices stored in the array In.
19 % 4. On line 362, default error tolerance for numerical
20 % integration can be changed.
21
22 %% Primary Function
23
24 function [L1,L2,V1,V2,A,D,OM,G1,G2,T1,T2,S,po,Dpo,p,Dp,g,q] =...
25                            SystemDecoupling(M,C,K,dt,t,qo,Dqo,f)
26
27 % SystemDecoupling calls five subfunctions and runs three
28 % automatic checks
29
30 % Automatic check 1: CheckMass determines if mass matrix M is
31 % nonsingular. If M is singular, program stops and output is
32 % set to zero
33 n = length(M); % Compute the number of degrees of freedom
34 if abs(det(M)) < (n^2)*(1e-4)
35     if det(M) == 0
36         disp('Program stops: mass matrix is singular')
37     else
38         disp('Program stops: mass matrix is almost singular')
39     end % Set output to zero
40         D=0; OM=0; g=0*qo*t; po=0; Dpo=0; T1=0; T2=0; V1=0; V2=0;
41         L1=0; L2=0; p=0*qo*t; Dp=0*qo*t; q = 0*qo*t; A=0; G1=0;
42         G2=0; S=0;
43     return
44 end % End of CheckMass
45
46 % Automatic check 2: check input parameters M, C, K, qo, Dqo,
47 % and f. If dimensions of input parameters are not compatible,
48 % program stops and output is set to zero
49 Sizes1 = [size(M),size(C),size(K),size(qo,1),...
50                                   size(Dqo,1),size(f,1)].';
51 Sizes2 = [size(qo,2),size(Dqo,2)].';
52 Sizes3 = size(f,2);
53 if any(Sizes1~=n) || any(Sizes2~=1) ||...
54                                  (Sizes3>1&&Sizes3<length(t))
55     disp('Program stops: input parameters are not conformable')
56     D=0; OM=0; g=0*qo*t; po=0; Dpo=0; T1=0; T2=0; V1=0; V2=0;
57     L1=0 ;L2=0; p=0*qo*t; Dp=0*qo*t; q = 0*qo*t; A=0; G1=0;
58     G2=0; S=0;
59     return
60 end
```

```

% Format user-supplied excitation: if t = 0:dt:T is a vector
% of m time points, then f is an nxm matrix and each column is
% the force at a time point. An input such as f = [1;2] is
% converted into an nxm array
temp = zeros(n,length(t));
for i = 1:n
    if length(f(i,:)) ~= length(t) && ...
                                    length(unique(f(i,:))) == 1
       temp(i,:) = f(i,1)*ones(1,length(t));
    else
        temp(i,:) = f(i,:);
    end
end
f = temp;
clear temp

% Decouple user-supplied system when all eigenvalues are distinct
    [A,D,OM,G1,G2,g,po,Dpo,T1,T2,S,CheckEigen,V1,V2,L1,L2] =...
                            SIAMdecoupling(M,C,K,qo,Dqo,f,dt);

% Find solution p(t) of decoupled system by direct simulation
if CheckEigen % All eigenvalues are distinct
    [tp,p,Dp] = simulate(eye(n),D,OM,g,[po;Dpo],t);
    p = real(interp1(tp,p',t)');   % Interpolate array to
                                   % conform with t
    Dp = real(interp1(tp,Dp',t)'); % Interpolate array to
                                   % conform with t

% Generate solution q(t) of user-supplied system by
% decoupling using p(t)
    q = T1*p + T2*Dp - T2*G1*f;

else % Some eigenvalues are repeated: output is set to zero
    p = 0*qo*t; Dp = 0*qo*t; q = 0*qo*t;
end
end % End of SystemDecoupling

%% Subfunctions

% There are four subfunctions

% -------------------------------------------------------------
% Subfunction 1: SIAMdecoupling is an implementation of
% decoupling by phase synchronization based on the algorithm
% depicted in Fig. 3.2

function [A,D,OM,G1,G2,g,po,Dpo,T1,T2,S,CheckEigen,V1,V2,L1,...
                        L2] = SIAMdecoupling(M,C,K,qo,Dqo,f,dt)

n = length(M); % Number of degrees of freedom
I = eye(n);    % Identity matrix
O = zeros(n);  % Matrix of zeros

```

```
% Solve quadratic eigenvalue problem
[V,L,CheckEigen] = solveQEP(M,C,K);

if ~CheckEigen % Some eigenvalues are repeated: output is set
               % to zero
    D=0; OM=0; g=0*f; po=0; Dpo=0; T1=0; T2=0; V1=0; V2=0;
    L1=0 ;L2=0; A=0; G1=0; G2=0; S=0;
else % All eigenvalues are distinct

%% Normalization of Eigenvectors

% After the eigensolutions have been indexed, the 2n
% eigenvectors are normalized in accordance with Eqs. (B.1)
% and (B.2) in Appendix B:
%   2*lambdaj*vj.'*M*vj + vj.'*C*vj = lambdaj - lambda(n+j)
%   2*lambda(n+j)*v(n+j).'*M*v(n+j) + v(n+j).'*C*v(n+j)...
%                                   = lambda(n+j) - lambdaj
% for 1 <= j <= n. The above normalization reduces to
% mass-normalization for undamped or classically damped
% symmetric-coefficient systems. This normalization will not
% be enforced if any scaling factor is greater than 1e4 in
% magnitude or if any scaling factor associated with a real
% eigenvector is imaginary. A different scheme for
% normalizing the eigenvectors may also be used.

switch_normalize = true; % Line 140
% Write true = normalize, false = do not normalize
if switch_normalize % Proceed to normalize if true
    str1 = 'Eigenvectors are not normalized because of';
    str2 = ' an imaginary scaling factor';
    str3 = ' a large scaling factor';
    msg1 = [str1 str2];
    msg2 = [str1 str3];
    s = zeros(2*n,1);
    for j = 1:n % Find all scaling factors
            s(j)   = 1/sqrt(2*L(j)*V(:,j).'*M*V(:,j) + ...
                           V(:,j).'*C*V(:,j))*sqrt(L(j) - L(j+n));
            s(j+n) = 1/sqrt(2*L(j+n)*V(:,j+n).'*M*V(:,j+n) + ...
                         V(:,j+n).'*C*V(:,j+n))*sqrt(L(j+n) - L(j));

            if round(imag(L(j)),6) == 0 && imag(s(j))~=0
                s = ones(2*n,1);
                disp(msg1)
                break
            end
            if round(imag(L(j+n)),6) == 0 && imag(s(j+n))~=0
                s = ones(2*n,1);
                disp(msg1)
                break
            end
            if  max(abs([s(j);s(j+n)])) > 1e4
                s = ones(2*n,1);
                disp(msg2)
                break
```

```
                end
        end

        for j = 1:n
            V(:,j)   = V(:,j)*s(j);
            V(:,j+n) = V(:,j+n)*s(j+n);

            if round(imag(L(j)),6) == 0
                if sum(round(V(:,j),6)<=0) == n
                    V(:,j) = -V(:,j); % Turn negative eigenvectors
                                      % positive
                end
                if sum(round(V(:,j+n),6)<=0) == n
                    V(:,j+n) = -V(:,j+n);
                end
            end
        end
        clear s
end % End of normalization of eigenvectors

% Enumerate eigenvalues and corresponding eigenvectors
    L1 = diag(L(1:n));
    L2 = diag(L(n+1:2*n));
    V1 = V(:,1:n);
    V2 = V(:,n+1:end);
    clear L V

% Find coefficients of decoupled system
    D  = - real(L1 + L2); % D is real
    OM = real(L1*L2);     % OM is real

% Find coefficients of excitation of decoupled system
    G1 = real((inv(V1*L1 - V2*L2/V2*V1)...
                    + inv(V2*L2 - V1*L1/V1*V2))/M);
    G2 = real((L1/(V1*L1 - V2*L2/V2*V1)...
                    + L2/(V2*L2 - V1*L1/V1*V2))/M);

% Find excitation g(t) of decoupled system
    g = (D*G1 + G2)*f + G1*gradient(f,dt);

% Find state companion matrix of user-supplied system
    A = [O I;-M\K -M\C];

% Find initial conditions of decoupled system
    P   = real([I I;L1 L2]*([V1 V2;V1*L1 V2*L2]\[qo;Dqo]));
    po  = P(1:n);
    Dpo = P(n+1:end) + G1*f(:,1);
    clear P

% Find coefficients of decoupling transformation in
% configuration space
    T1 = real((V1*L2 - V2*L1)/(L2 - L1)); % T1 is real
    T2 = real((V2 - V1)/(L2 - L1));       % T2 is real

```

```
% Find coefficient of decoupling transformation in state space
    S = real([I I;L1 L2]/[V1 V2;V1*L1 V2*L2]); % S is real
end
end % End of SIAMdecoupling

% ------------------------------------------------------------
% Subfunction 2: solveQEP solves quadratic eigenvalue problem
% and arranges the eigensolutions by a primary-secondary pairing
% scheme, as depicted in Task 1 of Chapter 3. The output V is an
% nx2n matrix of eigenvectors and L is an 2nx1 vector of
% eigenvalues

function [V,L,CheckEigen] = solveQEP(M,C,K)

n = length(M); % Number of degrees of freedom
I = eye(n);    % Identity matrix
O = zeros(n);  % Matrix of zeros

% Solution of quadratic eigenvalue problem by linearization
% in state space
[V,L] = eig([O I;-M\K -M\C]);
L = diag(L);

% Automatic check 3: CheckEigen determines if all eigenvalues
% are distinct. If some eigenvalues are equal, program stops and
% output is set to zero
CheckEigen = true;
if length(L) ~= length(unique(L)) % Condition for equal
                                  % eigenvalues
    CheckEigen = false;
    disp('Program stops: some eigenvalues are repeated')
    V = [O O]; L = zeros(2*n,1);
end % End of CheckEigen

if CheckEigen % All eigenvalues are distinct
% Specify tolerance for determining if eigenvalues are
% numerically close
    DistTol = 1e-4; % Line 260
% Two eigenvalues may be equal if their difference is less
% than DistTol. However, decoupling continues even when some
% eigenvalues are almost equal
    Dist = abs(L.' - L);    % Dist_ij = |lambda_i - lambda_j|
    Dist = triu(Dist);      % Take upper triangular part of
                            % symmetric matrix
    Dist = Dist(Dist ~= 0); % Convert to column array
    if min(Dist) < DistTol
        disp('Some eigenvalues may be repeated')
    end
clear DistTol Dist

%% Indexing of Eigensolutions

% The eigensolutions are arranged in accordance with a
% primary-secondary pairing scheme given by Eqs. (3.4)-(3.8).
```

```
% A different indexing scheme for the eigensolutions may be used.

% Find all complex eigenvalues and corresponding eigenvectors
    Inc = find(imag(L)~=0);
    LComplex = L(Inc);
    VComplex = V(:,Inc);

    Lc = LComplex(1:2:end); % First group of c complex
                            % eigenvalues
    Vc = VComplex(1:end/2,1:2:end);
    Lcn = LComplex(2:2:end); % Second group of c complex
                             % eigenvalues
    Vcn = VComplex(1:end/2,2:2:end);

% Indexing of complex eigensolutions: complex conjugates are
% always paired. Sort first c complex eigenvalues by increasing
% magnitude of imaginary parts
    [~,In] = sort(imag(Lc));
    Lc = Lc(In); Vc = Vc(:,In);
    Lcn = Lcn(In); Vcn = Vcn(:,In); % End of indexing of complex
                                    % eigensolutions

% Find all real eigenvalues and corresponding eigenvectors
    Inr = find(imag(L)==0);
    LReal = L(Inr);
    VReal = V(1:end/2,Inr);

% Indexing of real eigensolutions: there are
% (2*r)!/((2^r)*(r!)) different ways to pair the real
% eigensolutions. The 2r real eigenvalues are stored in
% LReal = [lambda_1;lambda_2;...;lambda_2r] in an order
% determined by the solution of eig(A). The indices are
% assigned by In = [1;2;...;2r]. The first r eigenvalues
% are matched with the remaining r eigenvalues to produce
% the pairing (lambda_j,lambda_(r+j)). Any pairing scheme
% consists of shuffling the indices stored in the array In.
% For example, the system with M = eye(2), C = [4 -1;-1 8],
% K = [1 0;0 4] has four real eigenvalues stored in
% LReal = [-7.7417;-3.4142;-0.2583;-0.5858]. One can use:

% In = randperm(length(LReal)); % Sort real eigenvalues randomly

% In = [2;1;3;4]; % Sort by manual specification
% The manually specified pairing scheme employs [2;1;3;4] for
% indexing of LReal and produces the pairing (-3.4142,-0.2583)
% and (-7.7417,-0.5858)

    switch_primary_secondary = true; % Line 324
% Write true = use, false = do not use
    if switch_primary_secondary % Use primary-secondary pairing
        [~,In] = sort(LReal); % Sort by increasing magnitude
    else
% Specify a pairing scheme if primary-secondary pairing is
% not used
```

```
        [~,In] = sort(LReal,'descend'); % Line 331
% Sort by decreasing magnitude, which produces a decoupled system
% equivalent to sorting the real eigenvalues by increasing
% magnitude
    end
    LReal = LReal(In);   % Arrange real eigenvalues after
                         % changing In
    VReal = VReal(:,In); % End of indexing of real eigensolutions

% Arrange all eigenvalues and corresponding eigenvectors
    L = [Lc; LReal(1:end/2); Lcn; LReal(end/2+1:end)];
    V = [Vc VReal(:,1:end/2) Vcn VReal(:,end/2+1:end)];
    clear Lc Vc Lcn Vcn LReal VReal Inc Inr In
end % End of indexing of eigensolutions
end % End of solveQEP

% -------------------------------------------------------------
% Subfunction 3: simulate computes the solution of a linear
% system by direct simulation using the solver ode45. The inital
% condition Qo is an 2nx1 vector with displacements on top of
% velocities

function [t,q,Dq] = simulate(M,C,K,f,Qo,t)

n = length(M);

% Construct state space equation
A = [zeros(n) eye(n);-M\K -M\C];
B = [zeros(n);M\eye(n)];

% Specify error tolerance for numerical integration
options = odeset('AbsTol', 1e-6, 'RelTol', 1e-4); % Line 362
% Absolute error tolerance has a default value of 1e-6 and
% relative error tolerance has a default value of 1e-4

Tspan = [0 t(end)]; % Time interval of integration
[t,Y] = ode45(@(T,y) SimResponse(T,y,t,f,A,B),Tspan,Qo,options);
q = Y(:,1:n)';
Dq = Y(:,n+1:end)';
end % End of simulate

% -------------------------------------------------------------
% Subfunction 4: SimResponse is a function handle to differential
% equations in state space for the solver ode45

function dydt = SimResponse(T,y,ft,f,A,B)

u = interp1(ft,f',T)';
dydt=A*y+B*u;
end % End of SimResponse
```

B.2 User Instructions

This program decouples any user-supplied linear system in real space with an extension of modal analysis. To be specific, let $\mathbf{A}_2$, $\mathbf{A}_1$, $\mathbf{A}_0$ arbitrary real square matrices (can be non-symmetric) of the same order with $\mathbf{A}_2$ assumed non-singular. A real and invertible transformation is generated to convert $\mathbf{A}_2\ddot{\mathbf{q}}+\mathbf{A}_1\dot{\mathbf{q}}+\mathbf{A}_0\mathbf{q} = \mathbf{f}(t)$ into $\ddot{\mathbf{p}} + \mathbf{D}\dot{\mathbf{p}} + \mathbf{\Omega}\mathbf{p} = \mathbf{g}(t)$ for which $\mathbf{D}$, $\mathbf{\Omega}$ are real and diagonal.

B.2.1 What Are the System Requirements?

This program runs on any (preferably 2017 or later) MATLAB platform. It consists of two files, `MAIN.m` and `SystemDecoupling.m`, with an overall size of about 18 KB. Only basic knowledge of MATLAB is required to use this program.

B.2.2 Are There Legal Restrictions on This Program?

The decoupling methodology is based upon original research of the developers and their collaborators. There are no legal restrictions for non-commercial use.

B.2.3 Are There Scientific Restrictions on This Program?

There are no scientific restrictions on the decoupling procedure. For numerical efficiency, it is assumed that all eigenvalues of $(\lambda^2\mathbf{A}_2+\lambda\mathbf{A}_1+\mathbf{A}_0)\mathbf{v} = \mathbf{0}$ are distinct.

B.2.4 How to Run the Program?

Place the two program files in the same folder. Open `MAIN.m` and supply 8 input parameters: $\mathbf{A}_2$, $\mathbf{A}_1$, $\mathbf{A}_0$, $\mathbf{f}(t)$, $\mathbf{q}(0)$, $\dot{\mathbf{q}}(0)$, T (suration of simulation), and dt (time increment in simulation). Run `MAIN.m`. It is not necessary to change the file `SystemDecoupling.m`.

B.2.5 What Is the Output?

Important output parameters include: (a) diagonal coefficients $\mathbf{D}$, $\boldsymbol{\Omega}$ and excitation $\mathbf{g}(t)$ of the decoupled system; (b) time histories of $\mathbf{p}(t)$, $\mathbf{q}(t)$; (c) decoupling transformations in both configuration and state spaces.

B.2.6 What Is the Normalization Scheme for Eigenvectors?

After the eigensolutions have been indexed, the $2n$ eigenvectors are normalized in accordance with

$$2\lambda_j \mathbf{v}_j^T \mathbf{A}_2 \mathbf{v}_j + \mathbf{v}_j^T \mathbf{A}_1 \mathbf{v}_j = \lambda_j - \lambda_{n+j} \tag{B.1}$$

and

$$2\lambda_{n+j} \mathbf{v}_{n+j}^T \mathbf{A}_2 \mathbf{v}_{n+j} + \mathbf{v}_{n+j}^T \mathbf{A}_1 \mathbf{v}_{n+j} = \lambda_{n+j} - \lambda_j \tag{B.2}$$

for $1 \leq j \leq n$. The above normalization reduces to mass-normalization for undamped or classically damped passive systems [4, 70]. This normalization maximizes the real parts of complex eigenvectors and minimizes their imaginary parts, thus allowing the real parts to be highlighted in analysis. This task is optional, and a different scheme for normalizing the eigenvectors may also be implemented.

B.2.7 Where to Give Suggestions or Get Support?

Contact Rubens G. Salsa Jr. at rsalsa@berkeley.edu.

B.2.8 Illustrative Examples

It will be shown how numerical results and analytical formulas are generated. Example B.1 highlights the unique capability of this program: decoupling in real space.

B.2.8.1 Example B.1

Consider the damped system governed by

$$\ddot{\mathbf{q}}+\begin{bmatrix}0.7 & -0.1\\ -0.1 & 0.2\end{bmatrix}\dot{\mathbf{q}}+\begin{bmatrix}2 & -1\\ -1 & 2\end{bmatrix}\mathbf{q}=\begin{bmatrix}\cos(t)\\ \sin(2t)\end{bmatrix} \tag{B.3}$$

and initial conditions $\mathbf{q}(0) = [1 \quad 2]^T$ and $\dot{\mathbf{q}}(0) = [-1 \quad 1]^T$. Since $\mathbf{CK} \neq \mathbf{KC}$, the system is non-classically damped and it cannot be decoupled by modal analysis. The associated homogeneous system was decoupled by phase synchronization in Example 2.4.

To use this program, choose, for convenience, $T = 20$ as the duration of simulation and $dt = 0.01$ as the time increment used in the simulation. Run `MAIN.m` with the following 8 input parameters:

```
M = eye(2);
C = [0.7 -0.1;-0.1 0.2];
K = [2 -1;-1 2];
qo = [1;2];
Dqo = [-1;1];
T = 20;
dt = 0.01;
f = [cos(t);sin(2*t)];
```

Run `MAIN.m`. To interpret the results, either type the parameter names in MATLAB command window or plot the output arrays. For example, the decoupled system is $\ddot{\mathbf{p}} + \mathbf{D}\dot{\mathbf{p}} + \boldsymbol{\Omega}\mathbf{p} = \mathbf{g}(t)$. The coefficient matrices can be obtained by typing `D` and `OM` in MATLAB command window:

```
>> D

D =

    0.3584         0
         0    0.5416

>> OM

OM =

    1.0337         0
         0    2.9022
```

An analytical formula for the excitation $\mathbf{g}(t)$ of the decoupled system can be constructed from

$$\mathbf{g}(t) = \left(\mathbf{D} + \mathbf{I}\frac{d}{dt}\right)\mathbf{G}_1\mathbf{f}(t) + \mathbf{G}_2\mathbf{f}(t). \tag{B.4}$$

To that end, retrieve the output parameters by typing `G1` and `G2` in MATLAB command window:

```
>> G1

G1 =

   -0.0948    0.0944
    0.0972    0.0952

>> G2

G2 =

    0.7498    0.7011
    0.6889   -0.7376
```

Type `po` and `Dpo` in MATLAB command window to obtain the initial conditions of the decoupled system as

```
>> po

po =

    2.3221
   -0.7111

>> Dpo

Dpo =

   -0.4267
   -1.6149
```

The configuration-space decoupling transformation is

$$\mathbf{q}(t) = \left(\mathbf{T}_1 + \mathbf{T}_2\frac{d}{dt}\right)\mathbf{p}(t) - \mathbf{T}_2\mathbf{G}_1\mathbf{f}(t). \tag{B.5}$$

The coefficients can be retrieved by typing

```
>> T1

T1 =

    0.7158    0.7415
    0.7349   -0.6860

T2

>> T2 =

   -0.0948    0.0972
    0.0944    0.0952
```

Finally, the state-space form of the decoupling transformation

$$\begin{bmatrix} \mathbf{p}(t) \\ \dot{\mathbf{p}}(t) \end{bmatrix} = \mathbf{S} \begin{bmatrix} \mathbf{q}(t) \\ \dot{\mathbf{q}}(t) \end{bmatrix} + \begin{bmatrix} \mathbf{0} \\ \mathbf{G}_1 \mathbf{f}(t) \end{bmatrix} \tag{B.6}$$

is obtained by typing

```
>> S

S =

    0.6740    0.7294   -0.0948    0.0944
    0.7474   -0.7282    0.0972    0.0952
    0.2840   -0.2836    0.7498    0.7011
   -0.0991   -0.0932    0.6889   -0.7376
```

B.2.8.2 Example B.2

Consider the linear system governed by

$$\ddot{\mathbf{q}} + \begin{bmatrix} 1 & 1 \\ -2 & 3 \end{bmatrix} \dot{\mathbf{q}} + \begin{bmatrix} -7 & 4 \\ -8 & 1 \end{bmatrix} \mathbf{q} = \mathbf{0} \tag{B.7}$$

and initial conditions $\mathbf{q}(0) = [1 \quad 2]^T$ and $\dot{\mathbf{q}}(0) = [-1 \quad 1]^T$. It was shown in Example 1.10 that this system can be decoupled by a similarity transformation into a system with complex coefficients. Phase synchronization decouples this system in real configuration space, as shown in Example 3.1. To decouple this system using this program, open `MAIN.m` and enter the following 8 input parameters:

```
M = eye(2);
C = [1,1;-2,3];
K = [-7,4;-8,1];
qo = [1;2];
Dqo = [-1;1];
T = 5;
dt = 0.01;
f = [0;0];
```

The decoupled system is given by $\ddot{\mathbf{p}} + \mathbf{D}\dot{\mathbf{p}} + \boldsymbol{\Omega}\mathbf{p} = \mathbf{0}$. Type `D` and `OM` in MATLAB command window:

```
>> D

D =
```

```
    6.1358         0
         0   -2.1358

>> OM

OM =

    9.4627         0
         0    2.6419
```

The coefficients $\mathbf{T}_1$ and $\mathbf{T}_2$ of the real configuration-space decoupling transformation

$$\mathbf{q}(t) = \left(\mathbf{T}_1 + \mathbf{T}_2 \frac{d}{dt}\right) \mathbf{p}(t) \tag{B.8}$$

is retrieved by typing

```
>> T1

T1 =

    2.6047   -0.2063
    1.2690    0.4241

T2

>> T2 =

    0.8763    0.3758
    0.5053    0.2167
```

The real initial conditions of the decoupled system are given by

```
>> po

po =

   -0.3679
    2.4633

>> Dpo

Dpo =

    1.3260
    3.4719
```

Finally, the state-space form of the decoupling transformation

$$\begin{bmatrix} \mathbf{p}(t) \\ \dot{\mathbf{p}}(t) \end{bmatrix} = \mathbf{S} \begin{bmatrix} \mathbf{q}(t) \\ \dot{\mathbf{q}}(t) \end{bmatrix} \tag{B.9}$$

is obtained by typing

```
>> S

S =

   -0.3212   -1.0516   -0.9946    1.0618
   -1.1995    1.3905   -0.4266    0.4554
    1.5317    2.9168    2.7970   -3.2422
    0.6569    1.2509    0.1377    0.4510
```

References

1. Rayleigh, J.W.S.: The Theory of Sound, vol. 1, reprint of the 1894 edn. Dover, New York (1945)
2. Meirovitch, L.: Analytical Methods in Vibrations. Macmillan, New York (1967)
3. Genta, G.: Vibration Dynamics and Control. Springer, New York (2009)
4. Chopra, A.K.: Dynamics of Structures: Theory and Applications to Earthquake Engineering, 5th edn. Pearson, Hoboken (2017)
5. Gawronski, W.K.: Dynamics and Control of Structures: A Modal Approach. Springer, New York (1998)
6. Humar, J.L.: Dynamics of Structures, 3rd edn. CRC Press, Boca Raton (2012)
7. Inman, D.J.: Vibration with Control, Measurement, and Stability. Prentice Hall, New Jersey (1989)
8. Clough, E.W., Penzien, J.: Dynamics of Structures, 2nd edn. McGraw-Hill, New York (1993)
9. Craig, R.R., Jr., Kurdila, A.J.: Fundamentals of Structural Dynamics, 2nd edn. Wiley, New York (2006)
10. Ewins, D.J.: Modal Testing: Theory and Practice, 2nd edn. Wiley, Philadelphia (2000)
11. He, J., Fu, Z.F.: Modal Analysis. Butterworth-Heinemann, Oxford (2001)
12. Roseau, M.: Vibrations in Mechanical Systems: Analytical Methods and Applications. Springer, New York (1987)
13. Weaver, W., Jr., Timoshenko, S.P., Young, D.H.: Vibration Problems in Engineering, 5th edn. Wiley, New York (1990)
14. Den Hartog, J.P.: Mechanical Vibrations, reprint of 1956 edn. Dover, New York (1985)
15. Newland, D.E.: Mechanical Vibration Analysis and Computation. Longman, Essex (1989)
16. Hurty, W.C., Rubinstein, M.F.: Dynamics of Structures. Prentice-Hall, Englewood Cliffs (1964)
17. Caughey, T.K., O'Kelly, M.E.J.: Classical normal modes in damped linear dynamic systems. ASME J. Appl. Mech. **32**(3), 583–588 (1965). https://doi.org/10.1115/1.3627262
18. Foss, K.A.: Co-ordinates which uncouple the equations of motion of damped linear dynamic systems. ASME J. Appl. Mech. **25**, 361–364 (1958)
19. Veletsos, A.S., Ventura, C.E.: Modal analysis of non-classically damped linear systems. Earthquake Eng. Struct. Dynam. **14**(2), 217–243 (1986). https://doi.org/10.1002/eqe.4290140205
20. Vigneron, F.R.: A natural modes model and modal identities for damped linear structures. ASME J. Appl. Mech. **53**(1), 33–38 (1986). https://doi.org/10.1115/1.3171734
21. Huseyin, K., Leipholz, H.H.E.: Divergence instability of multiple-parameter circulatory systems. Quart. Appl. Math. **31**(2), 185–197 (1973). https://doi.org/10.1090/qam/436698

R. Gonçalves Salsa Junior, F. Ma, *Advances in the Theory of System Decoupling*,
https://doi.org/10.1007/978-3-030-60846-0

22. Inman, D.J.: Dynamics of asymmetric nonconservative systems. ASME J. Appl. Mech. **50**(1), 199–203 (1983). https://doi.org/10.1115/1.3166991
23. Ahmadian, M., Inman, D.J.: Classical normal modes in asymmetric nonconservative dynamic systems. Am. Inst. Aeronaut. Astronaut. J. **22**(7), 1012–1015 (1984). https://doi.org/10.2514/3.48543
24. Caughey, T.K., Ma, F.: Complex modes and solvability of nonclassical linear systems. ASME J. Appl. Mech. **60**(1), 26–28 (1993). https://doi.org/10.1115/1.2900774
25. Lee, W.C., Ma, F.: Simultaneous triangularization of the coefficients of linear systems. ASME J. Appl. Mech. **64**(2), 430–432 (1997). https://doi.org/10.1115/1.2787328
26. Ma, F., Caughey, T.K.: Analysis of linear nonconservative vibrations. ASME J. Appl. Mech. **62**(3), 685–691 (1995). https://doi.org/10.1115/1.2896001
27. Ma, F., Imam, A., Morzfeld, M.: The decoupling of damped linear systems in oscillatory free vibration. J. Sound Vib. **324**(1–2), 408–428 (2009). https://doi.org/10.1016/j.jsv.2009.02.005
28. Ma, F., Imam, A., Morzfeld, M.: The decoupling of damped linear systems in free or forced vibration. J. Sound Vib. **329**(5), 3182–3202 (2010). https://doi.org/10.1016/j.jsv.2010.02.017
29. Morzfeld, M., Ma, F.: The decoupling of damped linear systems in configuration and state spaces. J. Sound Vib. **330**(2), 155–161 (2011). https://doi.org/10.1016/j.jsv.2010.09.005
30. Morzfeld, M., Ma, F., Parlett, B.N.: The transformation of second-order linear systems into independent equations. SIAM J. Appl. Math. **71**(4), 1026–1043 (2011). https://doi.org/10.1137/100818637
31. Kawano, D.T., Morzfeld, M., Ma, F.: The decoupling of defective linear dynamical systems in free motion. J. Sound Vib. **330**(21), 5165–5183 (2011). https://doi.org/10.1016/j.jsv.2011.05.013
32. Kawano, D.T., Salsa, R.G., Jr., Ma, F.: Decoupling of second-order linear systems by isospectral transformation. Zeitschrift für angewandte Mathematik und Physik **69**, 137 (2018). https://doi.org/10.1007/s00033-018-1030-x
33. Kawano, D.T., Morzfeld, M., Ma, F.: The decoupling of second-order linear systems with a singular mass matrix. J. Sound Vib. **332**(25), 6829–6846 (2013). https://doi.org/10.1016/j.jsv.2013.08.005
34. Garvey, S.D., Lancaster, P., Popov, A.A., Prells, U., Zaballa, I.: Filters connecting isospectral quadratic systems. Linear Algebra Appl. **438**(4), 1497–1516 (2013). https://doi.org/10.1016/j.laa.2011.03.040
35. Garvey, S.D., Friswell, M.I., Prells, U.: Co-ordinate transformations for second order systems. Part I: general transformations. J. Sound Vib. **258**(5), 885–909 (2002). https://doi.org/10.1006/jsvi.2002.5165
36. Garvey, S.D., Friswell, M.I., Prells, U.: Co-ordinate transformations for second order systems. Part II: elementary structure-preserving transformations. J. Sound Vib. **258**(5), 911–930 (2002). https://doi.org/10.1006/jsvi.2002.5166
37. Lancaster, P., Zaballa, I.: Diagonalizable quadratic eigenvalue problems. Mech. Syst. Signal Process. **23**(4), 1134–1144 (2009). https://doi.org/10.1016/j.ymssp.2008.11.007
38. Lanczos, C.: The Variational Principles of Mechanics, 4th edn. University of Toronto Press, Toronto (1970)
39. Greenwood, D.T.: Principles of Dynamics, 2nd edn., pp. 450–455. Prentice Hall, Englewood Cliffs (1988)
40. Müller, P.C., Schiehlen, W.O.: Linear Vibrations, pp. 23–29. Martinus Nijhoff Publishers, Dordrecht (1985)
41. Rosenberg, R.M.: Analytical Dynamics of Discrete Systems, pp. 221–222. Plenum Press, New York (1977)
42. Jordan, D.W., Smith, P.: Nonlinear Ordinary Differential Equations, 4th edn. Oxford University Press, Oxford (2007)
43. Strogatz, S.H.: Nonlinear Dynamics and Chaos. Perseus Books, Reading (1994)
44. Wiggins, S.: Introduction to Applied Nonlinear Dynamical Systems and Chaos, 2nd edn. Springer, New York (2003)

45. Brogan, W.L.: Modern Control Theory, 3rd edn., pp. 137–140. Prentice Hall, Upper Saddle River (1991)
46. Schmitz, P.D.: Normal mode solution to the equations of motion of a flexible airplane. J. Aircr. **10**(5), 318–320 (1973). https://doi.org/10.2514/3.44371
47. Soom, A., Kim, C.: Roughness-induced dynamic loading at dry and boundary-lubricated sliding contacts. J. Lubr. Technol. **105**(4), 514–517 (1983). https://doi.org/10.1115/1.3254655
48. Ma, F., Lee, W.C.: On the equations of non-linear vibrations. Int. J. Non-Linear Mech. **31**(6), 907–913 (1996). https://doi.org/10.1016/S0020-7462(96)00073-X
49. Fawzy, I., Bishop, R.E.D.: On the dynamics of linear non-conservative systems. Proc. R. Soc. Lond. A Math. Phys. Eng. Sci. **352**(1668), 25–40 (1976). https://doi.org/10.1098/rspa.1976.0161
50. Soong, T.T.: Active Structural Control: Theory and Practice. Longman, Essex (1990)
51. Hagedorn, P., DasGupta, A.: Vibrations and Waves in Continuous Mechanical Systems. Wiley, West Sussex (2007)
52. Grossmann, C., Roos, H.G., Stynes, M.: Numerical Treatment of Partial Differential Equations. Springer, New York (2007)
53. LeVeque, R.J.: Finite Difference Methods for Ordinary and Partial Differential Equations: Steady-State and Time-Dependent Problems. SIAM, Philadelphia (2007)
54. Johnson, C.: Numerical Solution of Partial Differential Equations by the Finite Element Method, reprint of the 1987 edn. Dover, New York (2012)
55. Zienkiewicz, O.C., Taylor, R.L., Zhu, J.Z.: The Finite Element Method: Its Basis and Fundamentals, 17th edn. Butterworth-Heinemann, Oxford (2013)
56. Lancaster, P., Tismenetsky, M.: The Theory of Matrices, 2nd edn. Academic Press, San Diego (1985)
57. Lancaster, P.: Lambda-Matrices and Vibrating Systems, reprint of the 1966 edn. Dover, New York (2002)
58. Gohberg, I., Lancaster, P., Rodman, L.: Matrix Polynomials. Academic Press, New York (1982)
59. Tisseur, F., Meerbergen, K.: The quadratic eigenvalue problem. SIAM Rev. **43**(2), 235–286 (2001). https://doi.org/10.1137/S0036144500381988
60. Trefethen, L.N., Bau, D., III, Fierro, R.D.: Numerical Linear Algebra. SIAM, Philadelphia (1998)
61. Demmel, J.W.: Applied Numerical Linear Algebra. SIAM, Philadelphia (1997)
62. Prells, U., Friswell, M.I.: A relationship between defective systems and unit-rank modifications of classical damping. ASME J. Vib. Acoust. **122**(2), 180–183 (2000). https://doi.org/10.1115/1.568458
63. Friswell, M.I., Prells, U., Garvey, S.D.: Low-rank damping modifications and defective systems. J. Sound Vib. **279**(3–5), 757–774 (2005). https://doi.org/10.1016/j.jsv.2003.11.042
64. Adhikari, S.: Damping modelling using generalized proportional damping. J. Sound Vib. **293**(1–2), 156–170 (2006). https://doi.org/10.1016/j.jsv.2005.09.034
65. Morzfeld, M., Ajavakom, N., Ma, F.: Diagonal dominance of damping and the decoupling approximation in linear vibratory systems. J. Sound Vib. **320**(1–2), 406–420 (2009). https://doi.org/10.1016/j.jsv.2008.07.025
66. Clough, R.W., Mojtahedi, S.: Earthquake response analysis considering non-proportional damping. Earthq. Eng. Struct. Dyn. **4**(5), 489–496 (1976). https://doi.org/10.1002/eqe.4290040506
67. Tsai, H.C., Kelly, J.M.: Non-classical damping in dynamic analysis of base-isolated structures with internal equipment. Earthq. Eng. Struct. Dyn. **16**(1), 29–43 (1988). https://doi.org/10.1002/eqe.4290160104
68. Tsai, H.C., Kelly, J.M.: Seismic response of the superstructure and attached equipment in a base-isolated building. Earthq. Eng. Struct. Dyn. **18**(4), 551–564 (1989). https://doi.org/10.1002/eqe.4290180409
69. Itoh, T.: Damped vibration mode superposition method for dynamic response analysis. Earthq. Eng. Struct. Dyn. **2**(1), 47–57 (1973). https://doi.org/10.1002/eqe.4290020105

70. Sestieri, A., Ibrahm, S.R.: Analysis of errors and approximations in the use of modal coordinates. J. Sound Vib. **177**(2), 145–157 (1994). https://doi.org/10.1006/jsvi.1994.1424
71. Horn, R.A., Johnson, C.R.: Matrix Analysis, 2nd edn. Cambridge University Press, New York (2013)
72. Gantmacher, F.R.: The Theory of Matrices, vol. 1. Chlesea Publishing Company, New York (1960)
73. Golan, J.S.: The Linear Algebra A Beginning Graduate Student Ought To Know, 3rd edn. Springer, New York (2012)
74. Taussky, O., Zassenhaus, H.: On the similarity transformation between a matrix and its transpose. Pac. J. Math. **9**(3), 893–896 (1959). https://doi.org/10.2140/pjm.1959.9.893
75. Ma, F., Caughey, T.K.: On the stability of linear and non-linear stochastic transformations. Int. J. Control **34**(3), 501–511 (1981). https://doi.org/10.1080/00207178108922544
76. Sagle, A.A., Walde, R.: Introduction to Lie Groups and Lie Algebra. Academic Press, New York (1973)
77. Freudenthal, H., De Vries, H.: Linear Lie Groups. Academic Press, New York (1969)
78. Zurmühl, R., Falk, S.: Matrizen und ihre Anwendungen, vol. 1, 5th edn. Springer, Berlin (1984)
79. Woodhouse, J.: Linear damping models for structural vibration. J. Sound Vib. **215**(3), 547–569 (1998). https://doi.org/10.1006/jsvi.1998.1709
80. Garvey, S.D., Prells, U., Friswell, M.I., Chen, Z.: General isospectral flows for linear dynamic systems. Linear Algebra Appl. **385**, 335–368 (2004). https://doi.org/10.1016/j.laa.2003.12.027
81. Prells U., Lancaster P.: Isospectral vibrating systems, part 2: structure preserving transformations. In: Langer M., Luger A., Woracek H. (eds) Operator Theory and Indefinite Inner Product Spaces. Operator Theory: Advances and Applications, vol. 163, pp. 275–298. Birkhäuser Verlag, Basel (2006). https://doi.org/10.1007/3-7643-7516-7_12
82. Kawano, D.T., Salsa, R.G., Jr., Ma, F., Morzfeld, M.: A canonical form of the equation of motion of linear dynamical systems. Proc. R. Soc. Lond. A Math. Phys. Eng. Sci. **471**, 20170809 (2018). https://doi.org/10.1098/rspa.2017.0809
83. Anaya, J.C.Z.: Diagonalization of quadratic matrix polynomials. Syst. Control Lett. **59**(2), 105–113 (2010). https://doi.org/10.1016/j.sysconle.2009.12.005
84. Morzfeld, M., Kawano, D.T., Ma, F.: Characterization of damped linear dynamical systems in free motion. Numer. Algebra Control Optim. **3**(1), 49–62 (2013). https://doi.org/10.3934/naco.2013.3.49
85. Gradshteyn, I.S., Ryzhik, I.M.: Table of Integrals, Series, and Products, 17th edn., p. 1060. Academic Press, San Diego (2007)
86. Duffin, R.L., Duffin, R.J.: A minimax theory for overdamped networks. J. Rational Mech. Anal. **4**, 221–233 (1955). http://www.jstor.org/stable/24900361
87. Barkwell, L., Lancaster, P.: Overdamped and gyroscopic vibrating systems. ASME J. Appl. Mech. **59**(1), 176–181 (1992). https://doi.org/10.1115/1.2899425
88. Bulatović, R.M.: Non-oscillatory damped multi-degree-of-freedom systems. Acta Mech. **151**(3), 235–244 (2001). https://doi.org/10.1007/BF01246920
89. Bulatović, R.M.: On the heavily damped response in viscously damped dynamic systems. ASME J. Appl. Mech. **71**(1), 131–134 (2004). https://doi.org/10.1115/1.1629108
90. Bhaskar, A.: Criticality of damping in multi-degree-of-freedom systems. ASME J. Appl. Mech. **64**(2), 387–393 (1997). https://doi.org/10.1115/1.2787320
91. Inman, D.J., Andry, A.N., Jr.: Some results on the nature of eigenvalues of discrete damped linear systems. ASME J. Appl. Mech. **47**(4), 927–930 (1980). https://doi.org/10.1115/1.3153815
92. Müller, P.C.: Oscillatory damped linear systems. Mech. Res. Commun. **6**(2), 81–85 (1979). https://doi.org/10.1016/0093-6413(79)90017-X
93. Nicholson, D.W.: Overdamping of a linear mechanical system. Mech. Res. Commun. **10**(2), 67–76 (1983). https://doi.org/10.1016/0093-6413(83)90075-7

94. Nicholson, D.W.: Eigenvalue bounds for linear mechanical systems with nonmodal damping. Mech. Res. Commun. **14**(2), 115–122 (1987). https://doi.org/10.1016/0093-6413(87)90028-0
95. Silvester, J.R.: Determinants of block matrices. Math. Gaz. **84**(501), 460–467 (2000). https://doi.org/10.2307/3620776
96. Ince, L.E.: Ordinary Differential Equations, pp. 394–395. Dover, New York (1956)
97. Zwillinger, D.: Handbook of Differential Equations, pp. 95–96. Academic Press, San Diego (1989)
98. Hill, J.M.: Solution of Differential Equations by Means of One-Parameter Groups, pp. 42–43. Pitman Publishing, Boston (1982)
99. Xu, J.: A synthesis formulation of explicit damping matrix for non-classically damped systems. Nucl. Eng. Des. **227**(2), 125–132 (2004). https://doi.org/10.1016/j.nucengdes.2003.08.005
100. Thomson, W.T., Calkins, T., Caravani, P.: A numerical study of damping. Earthq. Eng. Struct. Dyn. **3**(1), 97–103 (1974). https://doi.org/10.1002/eqe.4290030108
101. Angeles, J., Ostrovskaya, S.: The proportional-damping matrix of arbitrarily damped linear mechanical systems. ASME J. Appl. Mech. **69**(5), 649–656 (2002). https://doi.org/10.1115/1.1483832
102. Felszeghy, S.F.: On uncoupling and solving the equations of motion of vibrating linear discrete systems. ASME J. Appl. Mech. **60**(2), 456–462 (1993). https://doi.org/10.1115/1.2900815
103. Shahruz, S.M., Ma, F.: Approximate decoupling of the equations of motion of linear underdamped systems. ASME J. Appl. Mech. **55**(3), 716–720 (1988). https://doi.org/10.1115/1.3125855
104. Shahruz, S.M.: Approximate decoupling of the equations of motion of damped linear systems. J. Sound Vib. **136**(1), 51–64 (1990). https://doi.org/10.1016/0022-460X(90)90937-U
105. Udwadia, F.E.: Inverse problem of Lagrangian mechanics for classically damped linear multi-degrees-of-freedom systems. ASME J. Appl. Mech. **83**(10), 104501 (2016). https://doi.org/10.1115/1.4034012
106. Salsa, R.G., Jr., Kawano, D.T., Ma, F., Leitmann, G.: The inverse problem of linear Lagrangian dynamics. ASME J. Appl. Mech. **85**(3), 031002 (2018). https://doi.org/10.1115/1.4038749
107. Pars, L.A.: A Treatise on Analytical Dynamics. Wiley, New York (1965)

Index

R. Gonçalves Salsa Junior, F. Ma, *Advances in the Theory of System Decoupling*,
https://doi.org/10.1007/978-3-030-60846-0

The manufacturer's authorised representative in the EU is Springer Nature Customer Service Centre GmbH, Europaplatz 3, 69115 Heidelberg, Germany. If you have any concerns regarding our products, please contact ProductSafety@springernature.com

Printed and bound by CPI Group (UK) Ltd, Croydon, CR0 4YY

30/09/2026

02225267-0005